ISLAND RIVERS

Fresh Water and Place in Oceania

ISLAND RIVERS

Fresh Water and Place in Oceania

EDITED BY JOHN R. WAGNER
AND JERRY K. JACKA

Australian
National
University

PRESS

ASIA-PACIFIC ENVIRONMENT MONOGRAPH 13

ANU PRESS

Published by ANU Press
The Australian National University
Acton ACT 2601, Australia
Email: anupress@anu.edu.au

Available to download for free at press.anu.edu.au

A catalogue record for this book is available from the National Library of Australia

ISBN (print): 9781760462161
ISBN (online): 9781760462178

WorldCat (print): 1040273014
WorldCat (online): 1040272968

DOI: 10.22459/IR.06.2018

Cover design and layout by ANU Press. Cover photograph by Eric K. Silverman.

Contents

Tables

Figures

Contributors

Edvard Hviding is a Professor of Anthropology and Founding Director of the Pacific Studies Research group at the University of Bergen.

Jerry K. Jacka is an Associate Professor in the Department of Anthropology at the University of Colorado in Boulder.

Alexander Mawyer is an Associate Professor in the Centre for Pacific Islands Studies at the University of Hawai'i in Mānoa and Editor of *The Contemporary Pacific: A Journal of Island Affairs.*

Carlos Mondragón is an Associate Professor of Anthropology and a member of El Centro de Estudios de Asia y África (Center for Asian and African Studies) at El Colegio de México.

Marama Muru-Lanning is a Senior Research Fellow and Director in the James Henare Māori Research Centre at the University of Auckland.

Eric K. Silverman, former Research Professor of Anthropology at Wheelock College in Boston, is currently a Scholar at the Women's Studies Research Center at Brandeis University, USA.

Eilin Holtan Torgersen is a PhD candidate in the Department of Social Anthropology at the University of Bergen.

John R. Wagner is an Associate Professor of Anthropology at the University of British Columbia Okanagan.

1. Introduction: River as Ethnographic Subject

JOHN R. WAGNER, JERRY K. JACKA, EDVARD HVIDING,
ALEXANDER MAWYER AND MARAMA MURU-LANNING

Our intention in this volume is to make an original and innovative contribution to the ethnographic record of Oceania while also contributing to global debates about rivers and fresh water. Given the historical tendency for human societies everywhere to situate themselves near rivers, lakes, streams and other sources of fresh, flowing water, one might expect to find abundant, detailed descriptions of rivers in the early ethnographic record. But this is not the case.

Before 1950, when anthropological publications focused specifically on fresh water, the topic was most often folkloric in nature, about water beings, water divining or water symbolism (e.g. Hongi 1894; Holmes 1898; Piddington 1930). References to rivers as geographic markers were also very common, as when Murray and Ray (1918) published their article on 'The People and Language between the Fly and Strickland Rivers, Papua', but articles that actually describe rivers are rare indeed. Arthur Hayes (1906) published a description of the people of Abyssinia in a book entitled *The Source of the Blue Nile*, but wrote almost nothing about the river itself or about the riverine adaptations of the people he described. A 1912 article by Emmons about the Kitselas of British Columbia is exceptional, by comparison to most of the literature of that period, by virtue of its detailed environmental and, at times, poetic description of the river and its people:

> Where the Skeena River breaks through the eastern barrier of the Coast Range in British Columbia, 75 miles from its mouth, the pent-up waters have cut a deep cañon upward of a mile and a half in length, impassable during the spring and summer freshets and fraught with danger at all seasons. It is the most justly dreaded inland waterway of the Northwest, for, aside from the tremendous force of the contracted river over an uneven rocky bottom, forming great swirls and riffles, the upper entrance is obstructed by two high, narrow, rock ridges that divide the waters, forming two narrow channels at all stages and a third at extreme high water ... The natives named this part of the river cañon Dsilasshoo, and those who lived here as Gitdsilasshoo, "people of the cañon," but this is now written officially Kitselas. (Emmons 1912: 467–8)

By the middle of the twentieth century, however, water—and rivers in particular—came to occupy a more prominent place in the anthropological canon, as evidenced by the work of Julian Steward and others who were interested in the relationship of water to agriculture and political organisation. Steward published a seminal article about the role of irrigation in the development of 'ancient civilisations' in China, Mesopotamia, Peru and Mesoamerica in 1949. In 1955, with contributions from Karl Wittfogel and other authors, he published *Irrigation Civilizations*, a work that pre-dated Wittfogel's subsequently more influential work on 'hydraulic societies' (Wittfogel 1956, 1957). Steward and Wittfogel used mainly historical rather than ethnographic evidence to construct their theories, but ethnographic studies of irrigation systems also became prominent in the 1950s and have increased every decade since then. Thus, by mid-century, the field we now call the 'anthropology of water' came to be focused mainly on the value of water as a political and economic 'resource' rather than on its mythological and symbolic qualities (e.g. Geertz 1972). The post-war economic boom of the 1950s, '60s and early '70s, fuelled in part by the construction of ever more and larger hydro-power dams and massive irrigation diversions, consolidated this focus. The even more rapid pace of globalisation, population growth, urbanisation and agricultural intensification since the 1970s has led to a further re-inscription of rivers and fresh water, not just as a 'resource', but as a 'scarce resource' and 'commodity' (Bakker 2003; Strang 2004; Mehta 2005; Wagner 2012).

Though Steward's and Wittfogel's early theorising was deeply flawed in many of its details, especially in its tendency towards over-generalisation and deterministic arguments, scholars with diverse disciplinary backgrounds continue to make use of their general hypothesis about 'hydraulic societies' (see Worster 1985). However, the most nuanced contemporary studies of water in relation to political and economic organisation tend to avoid over-generalisations and focus on a much wider variety of historical settings and issues (e.g. Davis 1996, 2013; Donahue and Johnston 1997; Mosse 2003; Strang 2004, 2009; Baviskar 2007; Lahiri-Dutt and Samanta 2013; Wagner 2013; Hastrup and Hastrup 2016). Climate change adds yet more intensity and urgency to the anthropological study of water (Crate and Nuttall 2009; Hastrup 2009, Hastrup and Skrydstrup 2013; Orlove 2009), and converges with studies that focus on water as a human right (Sultana and Loftus 2012; Hossen 2014) and water and health (Whiteford and Whiteford 2005).

Not surprisingly, most anthropological studies of water over the past few decades have been carried out by political ecologists, but the interest in water has now expanded well beyond the set of interests and perspectives that define that field. This has given rise to a rich, diverse and even chaotic body of work that is also increasingly interdisciplinary. Some of the themes that attracted the interest of early anthropologists, such as myth and symbolism, are attracting renewed interest, but the attraction is now less folkloric than epistemological and ontological (e.g. Carse 2010; Krause 2010; Edgeworth 2011; Helmreich 2009; Strang 2014, 2016; Hastrup and Hastrup 2016; see also the contributions to this volume). As Hastrup and Hastrup write in their introduction to *Waterworlds*, 'anthropology now finds itself at a moment in time when the field is literally wide open', where 'there is no given anthropological object, but multiple and composite objects', and where 'fluidity on all accounts is the order of the day' (2016: 2). They apply their comments not just to the anthropology of water but to anthropology as a whole in an 'anthropocene era'.

The chapters in this volume are organised around the notion that rivers have social lives and therefore merit thorough ethnographic investigation and description.[1] Water is social by virtue of the fact that it flows through

---

1   There has been a curious proliferation of book titles recently that attribute social lives to a whole host of animate and inanimate beings. The trend appears to have begun with Appadurai's (1986) *The Social Life of Things*, a study of commodities, but has more recently been applied to 'trees' (Rival 1998), 'information' (Brown and Duguid 2000), 'coffee' (Cowan 2005), 'numbers' (Urton 2010), 'science' (Hastrup 2012), 'water' (Wagner 2013) and 'climate change models' (Hastrup and Skrydstrup 2013).

and animates all aspects of human life, not just at the cellular level but as a 'total social fact' in the sense proposed by Orlove and Caton (2010; see also Krause and Strang 2016). Using the term 'social' in its broadest sense to include political institutions, economic behaviour and cultural and religious practices, it is readily apparent that water informs every dimension of human social behaviour and is essential to the ways in which we construct class, gender and kinship.[2]

If one begins with the premise that rivers are fundamentally social in nature, then it is possible to treat rivers as the 'subject' rather than the 'object' of investigation. We are not collectively espousing a post-humanist approach to water, or an approach that depends on the argument that water possesses agency, though some contributors to this volume (e.g. Wagner, Jacka and Torgersen) are sympathetic to one or both of these approaches. Arguments about the agency of water are based on several distinct lines of thought, most of which are consistent with, but not essential to, the idea of rivers as ethnographic subjects. Scholars who adopt a post-humanist approach to their treatment of the non-human, for instance, argue that agency should not be defined in ways that restrict it to human beings. To do so, they argue, reinforces an anthropocentric approach that has imperilled the world, through pollution, over-exploitation of non-renewable resources, loss of biodiversity, climate change and other anthropogenic impacts. Actor-network theorists are also strong advocates for an approach that assigns agency to the non-human, defining agency not in terms of human intentionality, but in terms of measurable impacts within material-semiotic networks that include non-human as well as human actors. Other scholars embrace an animistic worldview in which all of nature is imbued with spirit and therefore intentionality and agency as well.

A recent publication by Strang (2014), with responses from several other scholars, effectively summarises many of the arguments for and against the agency of things. In this introduction, we would like to draw attention to two lines of argument not fully developed in that volume. The first of these is based on the work of Tim Ingold, who argues that the idea of non-human agency, despite its merits, is inherently self-contradictory. Strang (2014: 135) notes that Ingold's approach is consistent with 'a theoretical shift towards less anthropocentric visions of human-environmental

---

2   See also Strang (2004: 3–6) on the 'essentiality' of water.

interactions' and 'a more egalitarian bioethic of relationality', but that he nevertheless rejects the notion that things possess agency. Ingold argues that the 'problem of agency' is one we have created for ourselves:

> [It is] born of the attempt to re-animate a world already rendered lifeless by an exclusive focus on the 'objectness' of things … It is indeed striking that the more theorists have to say about agency, the less they seem to have to say about life. To rewrite the life of things as the agency of objects is to effect a double reduction, of things to objects, and of life to agency. (Ingold 2010: 97)

Ingold's argument did not directly inform our collective approach to this volume but it is very consistent with our intent. Our goal is to treat rivers as subjects rather than objects, and to shed light on the ways in which human lives in Oceania are interwoven with the lives of rivers, the landforms through which they flow and the other species they sustain.

A second line of argument not discussed in Strang (2014), but pertinent to this volume, has been developed by Salmond in her account of how the Whanganui River in New Zealand came to be recognised in law as a 'living being'. The Whanganui River is one of the first rivers in the world—though surprisingly not the only one—to have attained this status (Salmond 2014: 285–6). This was accomplished through the negotiation of a deed of settlement between the Māori and the New Zealand Government, known in Māori as Ruruku Whakatupua. In conferring 'legal personhood' on the river, Whanganui Māori are able to speak for their river in the country's courts and to file lawsuits on its behalf when environmental protections are not upheld. This approach is a type of modern co-management, through which the rights and health of the river are upheld through shared decision-making involving local Māori and other resident communities. Concerning the divergent understandings of the river held by Māori and non-Māori, Salmond writes:

> In the Whanganui deed of settlement with the Crown, ancestral Māori and modernist framings are juxtaposed, despite being incommensurable in certain respects. Drawing upon divergent forms of order, participants in the process have sought to weave together a concerted approach toward the management of New Zealand's waterways. This interweaving avoids the need for a merging of horizons, a "theory of everything" in which only one reality is possible and only one set of assumptions about the world can prevail. (Salmond 2014: 285)

Despite the fact that Salmond never cites Ingold in her paper, she concludes by articulating a position that seems very consistent with his, and also with the work of Bennett (2010), whom she does cite:

> In the Pacific and elsewhere, it is likely that the most intransigent obstacles to solving environmental (and other) challenges lie at the level of presupposition. As Jane Bennett has argued, 'the image of dead or thoroughly instrumentalised matter feeds ... our earth-destroying fantasies of conquest and consumption' (Bennett 2010: ix–x). Thus, Homo hubris trumps Homo sapiens. In such a situation, despite scientific claims to the contrary, epistemological solutions will not work. Fundamental conceptions and forms of order will have to shift if more lasting, flourishing styles of living are to be found. Here, ontological styles in which matter has never been dead or separated from people may prove helpful. When Bennett urges us to 'picture an ontological field without unequivocal demarcations between human, animal, vegetable, or mineral' where 'all forces and flows are or can become lively [and] affective' (Bennett 2010: 116–17), I think of the nineteenth–century Māori philosopher Nepia Pohuhu, who said, 'All things unfold their nature (*tupu*), live (*ora*), have form (*āhua*), whether trees, stones, birds, reptiles, fish, quadrupeds or human beings' (Pohuhu in Smith 1913: 13). (Salmond 2014: 305)

Neither Pohuhu's statement nor Salmond's argument should be read as an endorsement of animism, an idea invented by Western scholars and often misapplied to Indigenous societies around the world (Ingold 2006). The argument, rather, is about finding a way to think about rivers that is ontologically inclusive but also practical in political, economic and legal terms.

Understanding water as a subject in its own right, and not merely as 'instrumentalised matter', reinforces the distinction made by Ivan Illich (1985) between water as H_2O and water as the substance of dreams. Addressing the citizens of Dallas, Texas, in 1984 about a proposal to create a downtown lake a dozen blocks in size, Illich begins by stating that: 'I shall refuse to assume that all waters may be reduced to H_2O' (1985: 4). Relying on Bachelard's (1999) *Water and Dreams*, which explores the primordial, archetypal and imaginative dimensions of water, Illich writes:

> Following dream waters upstream, the historian will learn to distinguish the vast register of their voices. As his ear is attuned to the music of deep waters, he will hear a discordant sound that is foreign to waters, that reverberates through the plumbing of modern cities. He will recognize that the H_2O which gurgles through Dallas plumbing is not water,

but a stuff which industrial society creates. He will recognize that the twentieth century has transmogrified water into a fluid with which archetypal waters cannot be mixed. With enough money and broad powers to condemn and evict, a group of architects could very well create out of this sewage a liquid monument that would meet their own aesthetic standards. But since archetypal waters are as antagonistic to this new "stuff" as they are to oil, I fear that contact with such liquid monumentality might make the souls of Dallas's children impermeable to the water of dreams. (Illich 1985: 7)

In a more recent publication, Raffles, writing about the Amazon, also draws on Bachelard (1999) in order to emphasise the profound intimacy of the human/water relationship. He does so by discussing Bachelard's argument that 'the language of waters is a direct poetic reality', that 'murmuring waters teach birds to sing', that there is 'a continuity between the speech of water and the speech of man' (Bachelard 1983: 15, quoted in Raffles 2002: 179). Raffles' discussion recalls the earlier ethnographic work of Feld who, in an article entitled 'Waterfalls of Song', draws our attention to the ways in which the Kaluli people of Papua New Guinea create songs that imitate the flow of water. Feld writes that singing 'takes listeners on a journey that flows along local waterways and through local lands', and that 'the flow of these poetic song paths is emotionally and physically linked to the sensual flow of the singing voice' (1996: 91).

Rich ethnographic descriptions of human/river interactions can be found in many publications that do not necessarily constitute river ethnographies in the sense that we use the term here. Raffles, for instance, includes detailed descriptions of the middle Amazon where he conducted his research, but his book is a 'natural history' focused on a comparison of different constructions of Amazonian 'nature' (Raffles 2002: 7–9). Goldman (1963) provides a detailed physical description of the Cuduiarí River, a tributary of the Amazon, in his ethnography of the Cubeo people, and a thorough account of its religious significance in a more recent (2004) publication. But the river remains a supporting actor in these accounts, never the actual subject of the authors' investigation. In another study situated in Amazonia, Harris describes the floodplain adaptation of the *caboclo* community of Parú. He provides a rich, ethnographic description of river-based activities and excellent detail about the ecological characteristics of the floodplain, but, as he states on the opening page of his introduction, his book is about 'identity and change' (Harris 2001: 7), not about the river itself.

But river ethnographies in the sense that we use the term here, though still small in number, are becoming more common. Krause's recent (2010) study of the Kemi River in Finland is a prime example. He organised his study around the river itself, rather than a specific community or issue, and focused on the ways in which diverse individuals and communities participated in three dominant river activities—fishing, transportation and hydro-power generation. One of his main goals was to answer the question of whether 'river dwellers' in this setting 'think like a river'. 'To what extent', he asks, 'do the river and its flow pervade the imagination of those whose livelihoods are connected to its waters or along its banks?' (Krause 2010: 14)

In her study of the Murray River in Australia, Jessica Weir focuses on the relationships to the river of Indigenous people who see it as the heartbeat of a 'sentient ecology' (Weir 2009: 50, citing Anderson 2000: 116). Settler culture descendants, by contrast, whom Weir refers to as 'the moderns', view the river mainly as an economic resource. She also notes that the Murray Lower Darling Rivers Indigenous Nations are now calling for Indigenous water allocations based on the principle of 'cultural flows', meaning the amount of water required for them to maintain traditional cultural practices in all seasons (Weir 2009: 119).

Jessica Barnes takes a still different approach in *Cultivating the Nile*, an ethnographic study that emphasises the complex political and technological processes that determine how and when water will be delivered to farmers' fields in Fayoum Province, Egypt. 'This is a story about water politics', she writes, 'that takes the water itself, and the many ways in which that water is manipulated, as its beginning' (Barnes 2014: x). But, in addition to her focus on the quotidian water experiences of Fayoum farmers, she also seeks to trace the international dynamics of the Nile Basin.

> It is through close ethnographic observation of how people interact with water on a day-to-day basis that I am able to access the politics of the everyday. But I do not limit the view to Egypt alone. Rather, I situate this detailed, on-the-ground analysis within the context of processes, institutions, and technologies operating outside the nation's borders, which also affect how the water of the Nile flows into and through Egypt. Thus my analysis moves back and forth across space, shifting the gaze from downstream to upstream, farmer to donor agency, local to global, international water conference to irrigation canal. (ibid.: x)

Dancing with the River, another recent book-length river ethnography by Lahiri-Dutt and Samanta, focuses on the 'chars' that are constantly appearing and disappearing in 'the shallow riverbeds in the lower Gangetic plains of deltaic Bengal'. 'Nomadic chars', sandy islands and riverbank accretions that come into being as the result of the combined forces of erosion and sedimentation, provide homes for 'river-gypsies', 'wandering peoples' who 'inhabit a nonlegal, illegible, ungoverned, and ungovernable space, like many other peoples in Asia who have been, in James C. Scott's concise phrase, extruded by coercive state-making' (Lahiri-Dutt and Samanta 2013: ix).

The river ethnography genre also includes numerous shorter journal articles and book chapters. Stephen Lansing, best known for his studies of Balinese irrigation systems (Lansing 1991), has co-authored a study of the Skokomish River in Washington State that compares the ways in which the Indigenous Skokomish people value the river to those of the City of Tacoma, which operates a hydro-power facility on the river (Lansing et al. 1998). Studies of the impact of hydro-electric dams on the relationship of Indigenous peoples to their rivers constitute perhaps the largest single body of literature in the river ethnography genre (e.g. Baviskar 1995; Ettenger 1998; Colombi 2005; Féaux de la Croix 2011; also Muru-Lanning, this volume). Also worth noting, because of its relevance to contributions in this volume by Silverman and Wagner, is a paper by Harrison (2004), which focuses on the ways in which the Manambu people of Papua New Guinea use changes in the course of the Sepik River as a mnemonic device for both remembering and forgetting significant forms of collective experience.

As the 'subjects' of our ethnographic enquiries, we have given rivers free reign in this volume to inspire our imaginations in ways that sometimes converge with, and other times diverge from, the recounted experiences of our interlocutors, the river people who are equally the subjects of our various accounts. Also, though water as subject defies classification as mere H_2O, scarce resource or commodity, we do not abandon our interest in water as resource, but rather attempt to enrich our accounts through an interweaving of different perspectives.

The rivers on which we mainly focus are, of course, only one form of fresh water, and are no more or less important in hydrological terms than lakes, aquifers, springs, glaciers, clouds and precipitation. It is also worth noting that excellent ethnographic studies have been published that feature lakes (e.g. Orlove 2002) and estuaries, where fresh and salt water mix (Griffith 1999; Muehlmann 2013). Rivers do possess some unique social and ecological characteristics: the world's largest rivers—the Amazon, Nile, Yangtze, Mississippi, Volga, Ganges and Mekong—drain large portions of entire continents, and typically flow through several countries and thousands of communities on their way to the sea. Each community along the way constructs and reconstructs its own particular sets of relationships with the river—economic, ecological, religious, aesthetic, gendered and differentiated by class, caste and kin group. And the much more numerous, smaller rivers of the world, such as those common throughout Oceania, give rise to similarly diverse and intimate forms of social relations. Smaller, low-elevation islands, mainly coral atolls, often do not have rivers *per se*, but do have flowing water in the form of small, perhaps seasonal creeks, subterranean water, aquifers and springs. Thus, although we use the term 'rivers' to refer to our subject of study, other forms of flowing fresh water are considered as well.

The Rivers of Oceania

While descriptions of the sea are prominent in countless studies of Pacific island peoples, many of whom live their lives on ocean and lagoon shores, and who use the maritime domain for subsistence, transport and commerce (e.g. Harding 1967; Hviding 1996; D'Arcy 2006; Hau'ofa 2008), fresh water—a fundamental resource for everyday life everywhere—has remained largely absent from accounts of Pacific island environments and social lives. Only recently have the water resources of the Pacific Islands become a subject of some interest with regard to their vulnerability to the effects of climate change (Jacka 2009; Lazrus 2009, 2012; Duncan 2011; Rudiak-Gould 2013). In the case of atolls, with their precarious freshwater lenses supporting both people and cultivated crops (particularly swamp taro), thereby allowing for human existence in an environment not otherwise supportive of habitation, the absence of a research focus on the roles of fresh water is truly striking (but see Robertson 2013, 2016). And generally, given that most Pacific Islanders are not atoll dwellers, but for the most part live close to rivers, one might assume that ethnographers would have given more attention to rivers and what they provide in both material and symbolic terms. It is a great surprise then that scholarly attention has not been given to people's lives with rivers, even in New Guinea, which is one of the world's largest islands, with distinctive groups settled along the coast, by the rivers and in the valleys, hills and mountains of the interior.[3]

In the high islands of the Pacific, including those in the Melanesian archipelagos that approach the scale of small oceanic continents, and in some cases have substantial inland populations, the rivers that flow down from the central mountain ranges are of sufficient cultural significance to warrant closer ethnographic attention. Throughout the high islands of the tropical Pacific (excluding Papua New Guinea), there are far fewer people living inland than there were in pre-colonial times, yet some islands still have extensive settlements in the interior and provide reminders of a Pacific past when inland groups interacted with coastal groups in many different ways. These interactions were often structured according to economic specialisation, but also involved warfare as well as exchange, with rivers as a major scene for such interactions (Ross 1973; Gewertz 1983; Pomponio 1992; Lipset 1997; Bayliss-Smith et al. 2003). Today, rivers still represent important dimensions of everyday life, since they afford spatial mobility in the island topography and easy access to island interiors and their resources. Their headwaters and hilly tributaries provide water supplies for taro irrigation, and they are sources for the seasonal harvest of prized migratory fish species such as mullets. Most immediately, for those Pacific Islanders who do not live with the vulnerable groundwater sources on atolls and raised coral islands, it is the large and small watercourses in the hilly landscape of volcanic islands that provide for robust and stable supplies of often clean, high-quality water for drinking and other household uses in situations of typically high rainfall.

However, as Hviding emphasises in Chapter 2, rivers and fresh water occupy a significant place in the cosmological as well as material lives of Pacific Islanders. The Solomon Islanders whom he describes consider sea water and fresh water as opposing principles, and organise a good deal of their social and religious lives around that opposition. In that setting, large rivers descend from the dark and densely forested terrain of inland peoples and are perceived as inhospitable by those who live on the coast. Other contributors to this volume, while not focusing explicitly on this opposition, note its presence in their own Pacific island settings, but also indicate that it is constructed quite differently between one setting

3   The notable exception is anthropological work done along the mighty Sepik River (e.g. Gewertz 1983; Harrison 2004), which is itself constitutive of large-scale regional systems of exchange.

and another. Ultimately, as several contributors emphasise, the mouths of rivers are where fresh and salt waters mingle, and where particularly rich and biodiverse aquatic environments come into being. Thus, while many of the inland peoples of large Melanesian islands can accurately be described as river people, like the Eastern Iatmul of the Sepik River in Papua New Guinea (Silverman, this volume), engagements with rivers elsewhere in Oceania do not define people as 'riverine' but are a more or less integral part of everyday life.

This volume is unique by virtue of its focus on the riverine adaptations of Pacific island peoples, and in that sense can be said to fill a gap in the ethnographic record. But the gap is much too large to be filled by a single collection of papers. We hope it will encourage other scholars of Oceania to pay more attention to rivers and fresh water in the future, not just as a contribution to Oceanic literature, but as a uniquely positioned body of knowledge with relevance to other river and river mouth peoples around the world. In addition to the previously mentioned studies of the economic, political and cosmological characteristics of human/river relations, this volume also demonstrates the value and need for studies that examine their ontological, toponymic and linguistic characteristics. Ontological approaches necessarily raise questions about the place of nature in culture, as well as the place of water. Cronon's (1992) significant insights into the role of disciplinary narratives regarding 'nature' in the shaping of research agendas, and in choices about the objects scrutinised within those agendas, seem as relevant to one discipline as to another, and are certainly worth tracking in an interdisciplinary shift towards ontology.

Toponymy, the study of place names, offers particularly rich insights for those interested in studying fresh water from an ontological perspective. Place names provide insights into the eco-social anchoring of identity and social memory in watery sites, while also shedding light on the diverse ways in which such sites are categorised on the basis of diverse sets of values and experiences—whether economic, religious, historic or aesthetic. Place names often serve as anchors for the traditional histories through which knowledge, identity and social practices are transferred from one generation to the next. The study of place names also provides an entry point for the comparative study of water 'languages'—that is to say, the lexicon of water-related terms that at some levels are fairly consistent across large language families such as the Austronesian family, widely distributed throughout Oceania, but are also very localised, rooted in the particular circumstances, history and adaptation of a place-based speech community. In addition to the themes and approaches already identified, several authors in this volume address the distinction between flowing and blocked or still waters. Jacka, for instance, writing about the Porgeran people of Papua New Guinea (in Chapter 5), notes that flowing and still waters are seen as opposites, and that still waters are associated with blockages and illness, while flowing water is associated with health. The theme of blockages is addressed most explicitly by Mawyer in his account (in Chapter 4) of lost springs on Mangareva Island in Eastern Polynesia. In that setting, springs are no longer located where local knowledge holders expect them to be, whether because of blockages to intergenerational flows of knowledge or blockages and rerouting of subterranean flows of water. The opposition between blocked and flowing waters, like that between sea water and fresh water, emerges, in fact, as one of the primary themes interwoven throughout the different chapters. Muru-Lanning does not use the term 'blockage' in Chapter 6, but describes a situation in which multiple dams have been built on the Waikato River in order to generate hydro-power, thus blocking the natural flow of a river the Māori consider to be their Tupuna Awa ('River Ancestor'). Some Māori now find themselves in the difficult position of having to decide between becoming shareholders in privatised power companies created through the neo-liberal policies of the national government, which could possibly threaten their customary rights to rivers, or reject shareholder status in order to retain their more traditional guardianship relation, which is bound up with a spiritual and moral duty of care.

It is not surprising that the concept of flow emerged as a recurrent theme throughout this volume, given its prominence in other publications on water. Edgeworth (2011) has written an entire monograph on the *Archaeology of Flow*, pointing out the numerous ways in which rivers leave traces of their 'entanglements' with past human societies and constitute a form of cultural artefact. 'Flowing waters', he writes, also provide 'models for understanding other kinds of landscape flows' (Edgeworth 2011: 136), a comment reminiscent of Ingold's use of the concept of 'material flows'. Krause (2010: 264) has proposed that we should 'reclaim' the term 'flow' from its dominant use by anthropologists to refer to processes of globalisation, and instead deploy it, metaphorically and materially, to describe 'a world constituted by movement' (see also Mazzullo and Ingold 2008: 34–7; Linton 2010). Many scholars find water 'good to think', as Strang (2014: 134) has emphasised, and the concept of flow appears central to an ever-proliferating number of theoretical approaches. It is probably too late, as Féaux de la Croix (2014: 98) argues in response to Krause, to limit the use of this trope in the manner he suggests, but she does see another possible reason to limit its use. '[The] imagery of flow works beautifully for talking about certain kinds of connections', she writes, but '[it] works less well for talking about disconnection, inequality, injustice' (ibid.: 99). Several authors in this volume find, however, that the concept of blockage—the reverse of flow—does provide a useful way of discussing disconnection.

In the Waikato case study (Chapter 6), blockages of water are associated with commodification, another important theme interwoven throughout the volume. Muru-Lanning, interestingly, discusses commodification in the terms proposed by Kopytoff (1986), who discusses the tendency for objects to move into and out of states of commodification, depending on their cultural, historical and economic context. He uses slaves as his main example, noting that people can be captured, transformed into a commodity through sale in a slave market, but later on returned to non-slave status, at least under some circumstances. Kopytoff's juxtaposition of slavery with commodification is suggestive of another line of reasoning, one he did not pursue but which may still be relevant to our topic. The development aspirations of Pacific Islanders have been described in many publications, in both quotidian and archetypical contexts.[4] Given water's long history as an instrument of development, it is not surprising that water resources figure prominently in the development processes with which Pacific Islanders are now grappling. In situations where people feel desperate for development and are willing to risk their lives to achieve it, do they not become slaves of another kind, slaves of modernity, captured by processes of commodification that alienate them from the physical environment in which they live and from the products of their own labour (Graeber 2011)? This is a prominent issue throughout the island Pacific and all authors in this volume address it in one way or another. As Illich (1985) writes, contemporary industrial uses of water create a substance that is quite fundamentally different from the wild, flowing, archetypal waters that humans recognise as ancestors or the abode of ancestors, as healing or mysterious, and as the very source of life.

---

4   For example, see Lindstrom (1993) and Kaplan (1995) on cargo cults, and Errington and Gewertz (2004) on 'Yali's question'.

Organisation of the Book

In Chapter 2, our first case study, Edvard Hviding describes the coastal peoples of New Georgia Island, one of the larger volcanic islands in the Solomon island chain. River and sea, fresh water and salt water are 'foundational opposites' in this setting, and associated with them are other oppositions, most notably that between crocodile and shark. For the saltwater people of Marovo Lagoon, the large muddy rivers of New Georgia Island are a 'hostile realm' leading away from an open sunlit shore and lagoon to a dark, unknown and mountainous interior. Hviding thus provides us with a classic coastal Pacific island perspective on rivers, one that is shared, though with considerable variation, across Oceania. Having established the historical nature and cosmological significance of these classic oppositions, he concludes with the description of a newly constructed piped water system that now carries water from a New Georgia river across a lagoon to a small coral island offshore, an innovation that confounds and reconfigures classic oppositions in interesting ways. The piped water system is a practical convenience but also serves as a marker of modernity in this setting. So, in this chapter, we begin our examination of the relationship of development and modernity in Oceania to human/water relations.

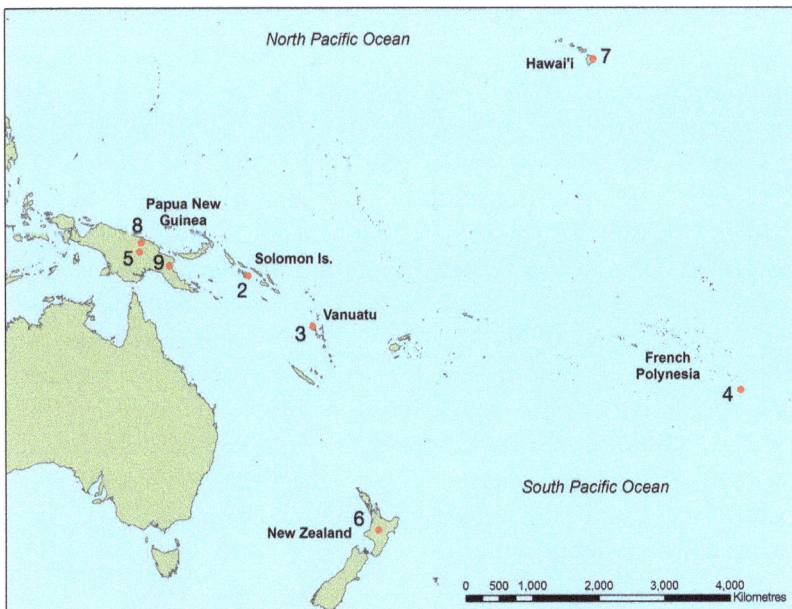

Figure 1.1 Map of the Pacific Ocean, showing locations by chapter.
Source: Cartography by Jerry Jacka.

In Chapter 3, Carlos Mondragón describes the layered historical meanings of the Jordan River on Espiritu Santo Island, the largest river in the Vanuatu archipelago. The Jordan drains northwards through a sparsely populated region of the island before emptying into Big Bay and the Pacific Ocean. Although he notes the opposition of 'bush' to coastline in this setting, the opposition is muted by comparison to that described by Hviding. However, a saltwater spring located near Big Bay gives rise to a small creek that flows into the bay and is considered 'opposite and complementary to the source of fresh water from which the River Jordan springs'. Mondragón also begins our discussion of freshwater toponymy, which is the study of place names in relation to the cultural identities of particular village sites and kin groups. In the North Santo setting he describes, place names reveal a rich, long-term though mutable set of historical relationships of particular kin groups to particular places and resources within the Jordan River watershed. Christian meanings, as indicated by the current English name of the river, simply add one more layer of meaning to an already complex tapestry.

In Chapter 4, Alex Mawyer also investigates the meanings of place names in his study of the Gambier Islands in Eastern Polynesia while introducing the reader to the richness of Polynesian and Austronesian vocabularies with respect to fresh water. On Mangareva, the largest of the Gambier Islands, certain springs are closely associated with myths and the historical practices of island chiefs. The intergenerational transmission of knowledge about these springs is being lost, however, and in many cases the springs themselves no longer exist in their former locations. Have the springs moved due to hydrological blockages of underground flows? Or is it just the intergenerational transmission of knowledge that has been blocked? Or perhaps both? Mawyer thus introduces us to one of the major themes that runs throughout the book, noting both the material and symbolic aspects of blockages and their contrast with perceptions of free-flowing water. He avoids easy generalisations about the association of blockages and other forms of human interference in water flows with particular historical events—whether the original peopling of Polynesia or the more recent push towards modernity. The theme of blockages proves to be a fertile one for the volume as a whole but one susceptible to varied applications and interpretations.

The theme of blockages is also prominent in Chapter 5, in Jerry Jacka's description of the impact of gold mining on the Porgera River in Papua New Guinea. The Porgeran people are located well inland, far from the coast, and here the main opposition is not between salt and fresh water but between moving and still waters, and more generally between movement and stasis. Movement of fluids in the body is associated with health, whereas blockages are associated with illness. Tailings from the Porgera gold mine are dumped directly, untreated, into the Porgera River, one of only four locations in the world where this is permitted, with disastrous results for local communities. Borrowing from Foster (2000), Jacka uses the concept of a metabolic rift to characterise the impact of the mine on Porgeran culture, kinship structures and livelihood strategies. Jacka also shows how rivers shape oppositional social identities as Porgerans and non-Porgerans struggle over access to mineral-rich lands.

In Chapter 6, Muru-Lanning describes the relationship of Māori communities in New Zealand to the Waikato River. This continues the theme of industrialisation begun by Jacka but in the context of hydro-power generation rather than mining. Once again, we are dealing with inland communities whose social structures and cosmology owe a great deal to the river along which they live, and Muru-Lanning provides a detailed description of these relations. In this case, however, the focus is on the ways in which commodification and privatisation have confounded traditional relationships. Traditionally, the Māori consider the Waikato, like the Whanganui, to be their Tupuna Awa, or River Ancestor, a relationship incompatible with the commodification of the river for hydro-electricity generation, and also incompatible with the system of European property rights imposed on the Māori through the much earlier Treaty of Waitangi. As in Chapter 3, we see that historical transformations of river relations involve the steady accretion and layering of meanings rather than the simple displacement of one set of meanings by another.

Eilin Torgersen, in Chapter 7, provides us with another classic example of how rivers throughout the world are being commodified for sale in global marketplaces, in this case through tourism rather than hydro-power or mining. 'Tourists come to Hawai'i', Torgersen writes, 'searching for the perfect waterfall that dives into a refreshing freshwater pool, surrounded by lush vegetation and beautiful flowers—a perfect scene for romance or adventurous cliff-diving activities'. Nevertheless, she continues, 'their dream of paradise is often shattered when the local population or their tour guide warns them about dangerous currents, sharp subsurface rocks and what a pounding 50-metre high waterfall will do to your head if you stand under it'. In her account of the Wailuku River that empties into the

Pacific Ocean at Hilo on the Big Island, she demonstrates just how much Indigenous ideas about the river differ from tourist representations, and argues that many native Hawaiians deliberately exaggerate the dangers associated with the rivers in order to limit tourist incursions. None of the other chapters engage with forms of impression management as directly as Torgersen, but her approach highlights an important theme that runs throughout the collection. The value and meaning of flowing waters is continually subject to negotiation and strategic deployment within and among communities.

None of the peoples described in this volume have a deeper historical or mythological attachment to their river than the Eastern Iatmul have to the Sepik River, as described by Eric Silverman in Chapter 8. The theme of economic development is as prominent in this chapter as in those that address the uses of rivers for hydro-power generation, tailings disposal and tourism, but in this case the relationship is inverted. The residents of Tambunum village have begun to view their river as an impediment rather than a pathway to development. Even though it is subject to large-scale flooding, it is unsuitable for hydro-development and is experiencing a diminishing tourist trade. 'Other regions of Papua New Guinea benefit from petroleum, natural gas, gold, and copper. But the people of Tambunum say that they "only have water and fish".' And even the fisheries are relatively unproductive. River identity, long a matter of pride for communities like Tambunum, is now a symbol of underdevelopment, backwardness and poverty. This chapter is especially telling in its depiction of a people's aspirations for modernity and the conflict between those aspirations and their historic appreciation of their river identity.

Wagner concludes the volume with a chapter that interprets landscape and knowledge transformations as processes of memory and forgetting. Four rivers dominate the social and economic lives of the people of Kamu Yali in Papua New Guinea, but the values and meanings associated with those rivers are subject to continual transformation. Kamu Yali residents selectively remember and forget specific practices, relationships and forms of knowledge about their rivers as they value them alternately as hunting and fishing resources, the ideal location for food gardens, as the source of building materials, fresh water or, most recently, as chromium mining sites. Wagner speaks to many of the same themes as the other contributors—to the ways in which kin group identities are constructed through relationships to specific rivers, to blockages, commodification, industrialisation and metabolic rifts, to the ways in which river relationships are transformed by the inexorable push towards modernity.

But he also provides a metanarrative for the volume as a whole—one that foregrounds the enduring significance of fresh and flowing water as a subject for contemporary ethnography and emphasises the mimetic relationship of culture to flowing water.

References

Anderson, D.G., 2000. *Identity and Ecology in Arctic Siberia: The Number One Reindeer Brigade*. Oxford: Oxford University Press.

Appadurai, A. (ed.), 1986. *The Social Life of Things: Commodities in Cultural Perspective*. Cambridge: Cambridge University Press. doi.org/10.1017/CBO9780511819582

Bachelard, G., 1999. *Water and Dreams: An Essay on the Imagination of Matter* (3rd edition, transl. E.R. Farrell). Dallas (TX): Dallas Institute of Humanities and Culture.

Bakker, K., 2003. *An Uncooperative Commodity: Privatizing Water in England and Wales*. Oxford: Oxford University Press.

Barnes, J., 2014. *Cultivating the Nile: The Everyday Politics of Water in Egypt*. Durham (NC): Duke University Press. doi.org/10.1215/9780822376217

Baviskar, A., 1995. *In the Belly of the River Tribal Conflicts over Development in the Narmada Valley*. Delhi: Oxford University Press.

—— (ed.), 2007. *Waterscapes: The Cultural Politics of a Natural Resource*. New Delhi: Permanent Black.

Bayliss-Smith, T., E. Hviding and T. Whitmore, 2003. 'Rainforest Composition and Histories of Human Disturbance in Solomon Islands.' *Ambio* 32: 346–352. doi.org/10.1579/0044-7447-32.5.346

Bennett, J., 2010. *Vibrant Matter: A Political Economy of Things*. Durham (NC): Duke University Press.

Brown, J.S. and P. Duguid, 2000. *The Social Life of Information*. Cambridge (MA): Harvard Business School Press.

Carse, A., 2010. 'Editor's Introduction.' *Cultural Anthropology* (virtual issue on 'Water'). Viewed 7 April 2017 at: production.culanth.org/curated_collections/10-water

Colombi, B.J., 2005. 'Dammed in Region Six: The Nez Perce Tribe, Agricultural Development, and the Inequality of Scale.' *American Indian Quarterly* 2: 560–589. doi.org/10.1353/aiq.2005.0085

Cowan, B., 2005. *The Social Life of Coffee: The Emergence of the British Coffeehouse.* New Haven (CT): Yale University Press.

Crate, S.A. and M. Nuttall (eds), 2009. *Anthropology and Climate Change: From Encounters to Actions.* Walnut Creek (CA): Left Coast Press.

Cronon, W., 1992. 'A Place for Stories: Nature, History, and Narrative.' *Journal of American History* 78: 1347–1376. doi.org/10.2307/2079346

D'Arcy, P., 2006. *The People of the Sea: Environment, Identity and History in Oceania.* Honolulu: University of Hawai'i Press.

Davis, W., 1996. *One River: Explorations and Discoveries in the Amazon Rainforest.* New York: Simon & Schuster.

——, 2013. *River Notes: A Natural and Human History of the Colorado.* Washington (DC): Island Press.

Donahue, J. and B.R. Johnston (eds), 1997. *Water, Culture and Power: Local Struggles in a Global Context.* Washington (DC): Island Press.

Duncan, D., 2011. 'Freshwater under Threat—Pacific Islands: Vulnerability Assessment of Freshwater Resources to Environmental Change.' Bangkok: United Nations Environment Programme.

Edgeworth, M., 2011. *Fluid Pasts: Archaeology of Flow.* London: Bloomsbury.

Emmons, G.T., 1912. 'The Kitselas of British Columbia.' *American Anthropologist* 14: 467–471. doi.org/10.1525/aa.1912.14.3.02a00030

Errington, F.K. and D.B. Gewertz, 2004. *Yali's Question: Sugar, Culture, and History.* Chicago: University of Chicago Press.

Ettenger, K., 1998. '"A River That Was Once So Strong and Deep": Local Reflections on the Eastmain Diversion, James Bay Hydroelectric Project.' In J. Donahue and B.R. Johnston (eds), *Water, Culture and Power: Local Struggles in a Global Context.* Washington (DC): Island Press.

Féaux de la Croix, J., 2011. 'Moving Metaphors We Live By: Water and Flow in the Social Sciences and around Hydroelectric Dams in Kyrgyzstan.' *Central Asian Survey* 30: 487–502. doi.org/10.1080/02634937.2011.614097

——, 2014. 'Everybody Loves Flow': On The Art Of Rescuing A Word From Its Users. *Suomen Antropologi* 39(2): 97–99.

Feld, S., 1996. 'Waterfalls of Song: An Acoustemology of Place Resounding in Bosavi, Papua New Guinea.' In S. Feld and K.H. Basso (eds), *Senses of Place*. Santa Fe (NM): School of Advanced Research Seminar Series.

Foster, J.B., 2000. *Marx's Ecology: Materialism and Nature*. New York: Monthly Review Press.

Geertz, C., 1972. 'The Wet and the Dry: Traditional Irrigation in Bali and Morocco.' *Human Ecology* 1: 23–39. doi.org/10.1007/BF01791279

Gewertz, D.B., 1983. *Sepik River Societies: A Historical Ethnography of the Chambri and Their Neighbors*. New Haven (CT): Yale University Press.

Goldman, I., 1963. *The Cubeo: Indians of the Northwest Amazon*. Urbana (IL): University of Illinois Press.

Graeber, D., 2011. *Debt: The First 5,000 Years*. Brooklyn (NY): Melville House.

Griffith, D., 1999. *The Estuary's Gift: An Atlantic Coast Cultural Biography*. University Park: Pennsylvania State University Press.

Harding, T.G., 1967. *Voyagers of the Vitiaz Strait: A Study of a New Guinea Trade System*. Seattle (WA): University of Washington Press.

Harris, M., 2001. *Life on the Amazon: The Anthropology of a Brazilian Peasant Village*. London: British Academy.

Harrison, S., 2004. 'Forgetful and Memorious Landscapes.' *Social Anthropology* 12: 135–151. doi.org/10.1017/S0964028204000436

Hastrup, K.B. (ed.), 2009. *The Question of Resilience: Social Responses to Climate Change*. Copenhagen: Danish Royal Society of Sciences and Letters.

——, 2012. 'The Social Life of Science.' *Public Service Review: Science and Technology* 15: 90–91.

Hastrup, K.B. and F. Hastrup (eds), 2016. *Waterworlds: Anthropology in a Fluid World*. New York (NY): Berghahn Books.

Hastrup, K.B. and M. Skrydstrup, 2013. *The Social Life of Climate Change Models: Anticipating Nature*. New York (NY): Routledge.

Hau'ofa, E., 2008. *We Are the Ocean: Selected Works*. Honolulu: University of Hawai'i Press.

Hayes, A.J., 1906. *The Source of the Blue Nile: A Record of a Journey Through the Soudan to Lake Tsana in Western Abyssinia, and of the Return to Egypt, by the Valley of the Atbara, with a Note on the Religion, Customs, &c. of Abyssinia*. London: Smith, Elder & Co.

Helmreich, S., 2009. *Alien Ocean: Anthropological Voyages in Microbial Seas.* Berkeley (CA): University of California Press.

Holmes, T.V., 1898. 'On the Evidence for the Efficacy of the Diviner and His Rod in the Search for Water.' *Journal of the Anthropological Institute of Great Britain and Ireland* 27: 233–259. doi.org/10.2307/2842965

Hongi, H., 1894. 'The Contest between Fire and Water.' *Journal of the Polynesian Society* 3: 155–158.

Hossen, A.M., 2014. Water Policy and Governance for the Empowerment of River Basin Communities in Rural Bangladesh. Okanagan: University of British Columbia (PhD thesis).

Hviding, E., 1996. *Guardians of Marovo Lagoon: Practice, Place, and Politics in Maritime Melanesia.* Honolulu: University of Hawai'i Press (Pacific Islands Monograph 14).

Illich, I., 1985. *H_2O and the Waters of Forgetfulness: Reflections on the Historicity of 'Stuff'.* Dallas (TX): Dallas Institute of Humanities and Culture.

Ingold, T., 2006. 'Rethinking the Animate, Re-animating Thought'. *Ethnos* 71(1): 9–20. doi.org/10.1080/00141840600603111

——, 2010. 'The Textility of Making.' *Cambridge Journal of Economics* 34: 91–102. doi.org/10.1093/cje/bep042

Jacka, J., 2009. 'Global Averages, Local Extremes: The Subtleties and Complexities of Climate Change in Papua New Guinea.' In S. Crate and M. Nuttall (eds), *Anthropology and Climate Change: From Encounters to Actions.* Walnut Creek (CA): Left Coast Press.

Kaplan, M., 1995. *Neither Cargo nor Cult: Ritual Politics and the Colonial Imagination in Fiji.* Durham (NC): Duke University Press. doi.org/10.1215/9780822381914

Kopytoff, I., 1986. 'The Cultural Biography of Things: Commoditisation as Process.' In A. Appadurai (ed.), *The Social Life of Things: Commodities in Cultural Perspective.* Cambridge: Cambridge University Press. doi.org/10.1017/CBO9780511819582.004

Krause, F., 2010. Thinking Like a River: An Anthropology of Water and its Uses along the Kemi River, Northern Finland. Aberdeen: University of Aberdeen (PhD thesis).

——, 2014. 'Reclaiming Flow for a Lively Anthropology.' *Suomen Antropologi* 39: 89–91.

Krause, F. and V. Strang, 2016. 'Thinking Relationships through Water.' *Society & Natural Resources* 29: 633–638. doi.org/10.1080/08941920.2016.1151714

Lahiri-Dutt, K. and G. Samanta, 2013. *Dancing with the River: People and Life on the Chars of South Asia*. New Haven (CT): Yale University Press. doi.org/10.12987/yale/9780300188301.001.0001

Lansing, J.S., 1991. *Priests and Programmers: Technologies of Power in the Engineered Landscape of Bali*. Princeton (NJ): Princeton University Press.

Lansing, J.S., P.S. Lansing and J. Erazo, 1998. 'The Value of a River.' *Journal of Political Ecology* 5: 1–22. doi.org/10.2458/v5i1.21395

Lazrus, H., 2009. 'The Governance of Vulnerability: Climate Change and Agency in Tuvalu, South Pacific'. In S. Crate and M. Nuttal (eds), *Anthropology and Climate Change: From Encounters to Actions*. Walnut Creek, CA: Left Coast Press.

——, 2012. 'Sea Change: Island Communities and Climate Change.' *Annual Review of Anthropology* 41: 285–301. doi.org/10.1146/annurev-anthro-092611-145730

Lindstrom, L., 1993. *Cargo Cult: Strange Stories of Desire from Melanesia and Beyond*. Honolulu: University of Hawai'i Press.

Linton, J., 2010. *What Is Water? The History of a Modern Distraction*. Vancouver: UBC Press.

Lipset, D., 1997. *Mangrove Man: Dialogics of Culture in the Sepik Estuary*. Cambridge: Cambridge University Press. doi.org/10.1017/CBO9781139166867

Mazzullo, N. and T. Ingold, 2008. 'Being Along: Place, Time and Movement among Sa´mi People.' In O.B. Jøren and B. Grana (eds), *Mobility and Place: Enacting Northern European Peripheries*. Aldershot: Ashgate.

Mehta, L., 2005. *The Politics and Poetics of Water: Naturalizing Scarcity in Western India*. Hyderabad: Orient Longman.

Mosse, D., 2003. *The Rule of Water: Statecraft, Ecology, and Collective Action in South India*. Oxford: Oxford University Press.

Muehlmann, S., 2013. *Where the River Ends: Contested Indigeneity in the Mexican Colorado Delta*. Durham (NC): Duke University Press. doi.org/10.1215/9780822378846

Murray, J.W.P. and S.H. Ray, 1918. 'The People and Language between the Fly and Strickland Rivers, Papua.' *Man* 18: 40–45. doi.org/10.2307/2788423

Orlove, B., 2002. *Lines in the Water: Nature and Culture at Lake Titicaca.* Berkeley: University of California Press.

——, 2009. 'The Past, the Present and Some Possible Futures of Adaptation.' In W.N. Adger, I. Lorenzoni and K. O'Brien (eds), *Adaptation to Climate Change: Thresholds, Values, Governance.* Cambridge: Cambridge University Press. doi.org/10.1017/CBO9780511596667

Orlove, B. and S.C. Caton, 2010. 'Water Sustainability: Anthropological Approaches and Prospects.' *Annual Review of Anthropology* 39: 401–415. doi.org/10.1146/annurev.anthro.012809.105045

Piddington, R., 1930. 'The Water-Serpent in Karadjeri Mythology.' *Oceania* 1: 352–354. doi.org/10.1002/j.1834-4461.1930.tb01656.x

Pomponio, A., 1992. *Seagulls Don't Fly into the Bush: Cultural Identity and Development in Melanesia.* Belmont (CA): Wadsworth.

Raffles, H., 2002. *In Amazonia: A Natural History.* Princeton (NJ): Princeton University Press.

Rival, L., 1998. *The Social Life of Trees: Anthropological Perspectives on Tree Symbolism.* Oxford: Berg.

Robertson, M.L., 2013. Connecting Worlds of Water: An Ethnography of Environmental Change on Tarawa, Kiribati. Copenhagen: University of Copenhagen (PhD thesis).

——, 2016. 'Enacting Groundwaters in Tarawa, Kiribati: Searching for Facts and Articulating Concerns.' In K. Hastrup and F. Hastrup (eds), *Waterworlds: Anthropology in Fluid Environments.* New York: Berghahn Books.

Ross, H.M., 1973. *Baegu: Social and Ecological Organization in Malaita, Solomon Islands.* Chicago: University of Illinois Press.

Rudiak-Gould, P., 2013. *Climate Change and Tradition in a Small Island State.* London: Routledge.

Salmond, A., 2014. 'Tears of Rangi: Water, Power, and People in New Zealand.' *HAU: Journal of Ethnographic Theory* 4: 285–309. doi.org/10.14318/hau4.3.017

Smith, S.P., 1913. *The Lore of the Whare Wananga* (2 volumes). New Plymouth (NZ): The Polynesian Society.

Steward, J.H., 1949. 'Cultural Causality and Law: A Trial Formulation of the Development of Early Civilizations.' *American Anthropologist* 51: 1–27. doi.org/10.1525/aa.1949.51.1.02a00020

Steward, J.H., R. McAdams, M.D. Collier, A. Palerm, K.A. Wittfogel and R. Beals, 1955. *Irrigation Civilizations: A Comparative Study.* Washington (DC): Pan American Union (Social Science Monograph 1).

Strang, V., 2004. *The Meaning of Water.* Oxford: Berg.

——, 2009. *Gardening the World: Agency, Identity and the Ownership of Water.* New York: Berghahn Books.

——, 2014. 'Fluid Consistencies: Material Relationality in Human Engagements.' *Archaeological Dialogues* 21: 133–150. doi.org/10.1017/S1380203814000130

——, 2016. 'Reflecting Nature: Water Beings in History and Imagination.' In K. Hastrup and F. Hastrup (eds), *Waterworlds: Anthropology in Fluid Environments.* New York (NY): Berghahn Books.

Sultana, F. and A. Loftus (eds), 2012. *The Right to Water: Politics, Governance and Social Struggles.* New York (NY): Earthscan.

Urton, G., 2010. *The Social Life of Numbers: A Quechua Ontology of Numbers and Philosophy of Arithmetic.* Austin (TX): University of Texas Press.

Wagner, J., 2012. 'Water and the Commons Imaginary.' *Current Anthropology* 53: 617–641. doi.org/10.1086/667622

—— (ed.), 2013. *The Social Life of Water.* New York (NY): Berghahn Books.

Weir, J.K., 2009. *Murray River Country: An Ecological Dialogue with Traditional Owners.* Canberra: Aboriginal Studies Press.

Whiteford, L. and S. Whiteford, 2005. *Globalization, Water and Health: Resource Management in Times of Scarcity.* Santa Fe (NM): School of American Research Press.

Wittfogel, K.A., 1956. *The Hydraulic Civilizations.* Chicago (IL): University of Chicago Press.

——, 1957. *Oriental Despotism: A Comparative Study of Total Power.* New Haven (CT): Yale University Press.

Worster, D., 1985. *Rivers of Empire: Water, Aridity and the Growth of the American West.* New York (NY): Pantheon.

2. The River, the Water and the Crocodile in Marovo Lagoon

EDVARD HVIDING

Introduction

In this chapter, I examine the interrelationships between salt and fresh water—ocean, lagoon and rivers—and land in the Marovo Lagoon area of the western Solomon Islands, based on my fieldwork there since 1986. The account is primarily an ethnographically descriptive one, grounded in what, during my first period of fieldwork in Marovo Lagoon (18 months, 1986–87), was a consistent immersion in everyday subsistence practice, involving daily participatory engagement with the environments of sea and land through fishing, gathering, hunting and gardening, and regular navigation of my own canoes around the lagoon and the rest of the New Georgia group of islands, of which the Marovo area forms the eastern part (Hviding 1996; see Figure 2.1). While I shall provide a general account of the cosmological foundations for the organisation of environmental and symbolic realms in New Georgia, in order to outline relationships between sea, land and river, I turn to the experience of everyday practice for ethnographic insights into how the people of Marovo Lagoon engage with their environments and the creatures they contain. In this regard, it is interesting that the relationships between people and fresh water, between land and river—so prominent until the late nineteenth century but later becoming less significant—have seen recent intensification through the environmental problems caused in the lagoon by large-scale logging operations on the land (carried out by Asian companies across New Georgia, in quite destructive fashion) and through the rapid proliferation of piped water supplies to villages.

Since the early twentieth century, just about all villages in New Georgia have been located in coastal areas, after demographic upheavals and transformations of political economy caused inland settlements and associated large-scale irrigated taro cultivation to be gradually abandoned (Hviding and Bayliss-Smith 2000: 145–52; Bayliss-Smith et al. 2003). The demographic pattern shown on present-day maps is, then, a historically specific configuration very different from that of pre-colonial times. Until the late nineteenth century, the flow of river water as a foundation for surplus-generating irrigation was central to the regional political dynamics of New Georgia. On this background, rivers are today repositories of complex inland histories, and river travel by those whose ancestral generations lived and cultivated in the interior lands of New Georgia can be a powerful statement about the longevity of territorial claims.

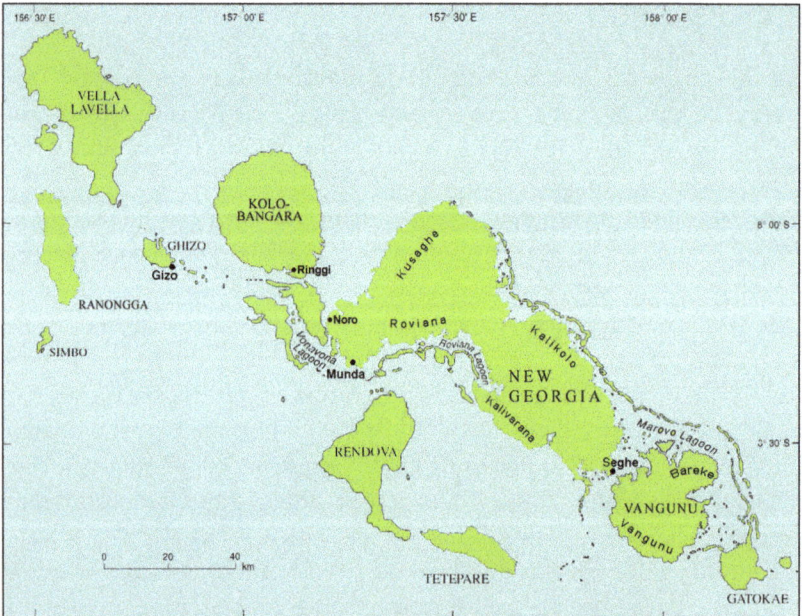

Figure 2.1 New Georgia and associated islands and lagoons.
Source: University of Bergen.

I shall approach river and water from a grounding in an inclusive sociality that, in everyday practice, connects human and non-human agents and domains on many levels of meaning. Building my discussion of water and rivers from an experience-based view of Melanesian social life, I take as my starting point some basic distinctions in the Marovo view of the lived, and living, world.

Sea, River, Territory: Templates of Existence in Marovo

In the Marovo language, salt and fresh water are clearly distinguished as *idere* and *kavo* respectively. Interestingly, while *idere* means sea or salt water, it also refers to fresh water that is hot (i.e. heated); whereas *kavo*, beyond being a generic term for fresh (drinkable) water, also refers to those major water courses classifiable as rivers. I shall build on this quite pliable distinction to provide some introductory ethnographic glimpses, first focusing on some key non-human species that have iconic value for the distinction.

It should come as no surprise that, in seaboard Melanesia, the most significant animal representatives of sea and river are two formidable predators that can be more than a match for the humans who encounter them: sharks (*kiso*) and crocodiles (*vua*). Since these predators have such potential to profoundly affect human existence, they are deeply tied into vernacular socialities through ancient association with specific kin groups whose members have, over the generations, engaged with sharks and crocodiles in their respective environments. At base, this has existential foundations in a historical distinction throughout the Marovo area (as in many other parts of island Melanesia) between 'bush people' and 'saltwater people', whereby the former have associations with the crocodile and the latter with the shark. Interestingly, the everyday habits of crocodiles to some degree mirror the historical pattern of everyday life among bush people, in terms of moving regularly between the inner lands and the coasts. And so, the specific environmental realm of rivers, connecting as it does the inner lands and the sea, is inextricably tied to the movements of that fearsome predator, the saltwater or estuarine crocodile (*Crocodylus porosus*). Conversely, the saltwater people's associations with sharks mirror the predatory maritime warfare that characterised the pre-colonial lives of those kin groups; the shark roams and ranges wide and far—in the lagoon, but also far beyond and 'up' to the open ocean.

The extraordinarily close maritime engagements of Marovo people (though mainly men) allow for detailed knowledge of many locally named species of sharks. A recent overview (Hviding 2005: 29–31) indicates that Marovo shark classification includes 15 or more vernacular taxa. However, the totemic traditions in question depart from such precise knowledge to focus simply on a generic, predatory ancestral shark that eats people (preferably enemies) and that moves freely, near and far, in maritime space, using the prominent topographic feature of passages through the barrier reef as its conduit between lagoon and ocean. In the Marovo scheme of spatial directions, the ancestral sharks move on a horizontal plane, *in* and *out* from open sea and lagoon. Unlike this open-ended mobility of the generic ancestral shark, the crocodile, of which there are only two Marovo taxa (one of which is that referred to in ancestral totemic traditions), maintains a shuttling behaviour back and forth along river courses, conceptualised in spatial schemes as vertical: *down* through estuaries into the lagoon and, from there, even all the way *up* to the barrier reef. These important cosmological foundations involving the shark and the crocodile amount to close descriptive accounts of the environmental dynamics of flow concerning sea and river.

The cyclical dynamics—tidal, lunar, seasonal—of the sea are elaborated in the Marovo concept of *kolokolo* ('time'), which is a reduplication of *kolo* ('ocean'). The notion of time itself thus appears to be modelled on a combination of oceanographic processes and human activities that are at once predictable, yet erratic, and observable, yet eternal and open-ended. Conversely, the river is a prime linear mover of land–sea relationships, both social and environmental. Although the tides flow and ebb in estuaries, and rivers may at high tide have salty surface water quite far inland, it is the powerful, insisting, unidirectional downwards movement of water from the inner lands to the sea that is the defining characteristic of rivers. I am going to discuss a range of past and present situations and contexts, generic and specific, in which this particular quality of rivers is at the forefront in Marovo and New Georgia more widely. While I rely on the Marovo language for vernacular concepts, I emphasise that those concepts from Marovo generally have cognates throughout the Austronesian languages of New Georgia, and my discussion is therefore of regional scale.

I have elsewhere analysed at length the everyday practice of saltwater people and the many dimensions of relationships in Marovo between people and sea (Hviding 1996), but the river as such is a long overdue focus in my writing and analysis concerning the western Solomons.

Hence, I now bring attention to the large rivers of the high volcanic island of New Georgia and its eastern neighbour Vangunu. These islands are two big, mountainous land masses extending from the northwest to the southeast for about 100 km, with the raised barrier reefs that define the large lagoons of Marovo and Roviana enclosing the northeastern and southwestern coasts of New Georgia, and the northern and eastern coasts of Vangunu (see Figures 2.1 and 2.2). The sheltered, island-dotted lagoons extending between the mainland coast and barrier reef islands are the central focus of the maritime lifestyle of the New Georgia people, who are renowned experts in fishing and marine gathering. With very few exceptions, villages of the present time are all in coastal locations, some on small low islands in the lagoon, or on the raised barrier reef itself.

Figure 2.2 Marovo Lagoon, showing rivers and place names.
Source: University of Bergen.

Transitional ecological zones give their own, particularly significant contributions to the resilient subsistence economy of the more than 30,000 villagers of the New Georgia Islands. While the ocean-facing reef walls of the barrier islands provide rich fishing grounds (particularly for underwater spearfishing), the extensive mangrove swamps that fringe the mainland coasts (and some lagoon-facing coasts of barrier islands) offer stocks of bivalve *Polymesoda* molluscs and mud crabs (*Scylla serrata*). As in many parts of the island Pacific, the fundamental New Georgian meal consists of some protein in the form of fish or shellfish, with the main ingredient being root crop staples and leafy vegetables obtained from shifting cultivation in the lower foothills behind the coastal village locations. But land and sea are interconnected realms in more than economic and culinary senses; they form an interrelated continuum in the pervasive Marovo concept of *puava*, a territory conceptualised much like the *vanua* of Fiji and the *ahupua'a* of Hawai'i, which ranges over a succession of ecological zones and is inhabited by the people who control it as their ancestral estate. The *puava* (referred to in other New Georgian languages by cognates such as *pepeso*, *pepeha*, etc.) is a cosmological foundation whereby social group and territory are mutually constituted, and is additionally significant in the New Georgian context of bilateral kinship, where it is shared belonging to, and engagement with, a named *puava* that defines the kin group (Hviding 2003a).

At its greatest conceptualised extent, a particular named *puava*, held collectively as ancestral estate by the recognised members of a homonymous cognatic kin group referred to as a *butubutu*, may extend from the mountain tops and high-elevation ridges in the interior, through mountain slopes, foothills and lowland basins to the mangrove-fringed coast, across the lagoon, out to and including the barrier reef. This pattern characterises the *puava* of 'mixed' *butubutu* that have formed historically from amalgamations of bush and saltwater groups, and as such are simultaneously coastal and bush (as well as river) people (see Wagner, this volume). It should be noted here that quantification of land-holding and sea-holding *butubutu* throughout the Marovo area is not a straightforward matter, owing to bilateral kinship and the consequent multiplicity of social relations, coupled with historical upheavals. A cautious estimate would imply about a third of the recently significant kin groups in each of the saltwater, bush and mixed categories, though with a somewhat smaller demographic proportion for saltwater groups (Hviding 1996: 377–81). As a result, most *butubutu* do remain of saltwater or bush orientation, and their *puava* include either coastal fringes with large areas of lagoon and barrier reef, and islands therein, or vast tracts of land extending from interior mountain tops all the way down to the lagoon shore. The rivers that are such prominent features, indeed makers, of the landscape (and seascape) are thus mainly under the control of *butubutu* considered to be of bush orientation and, in a very important sense, remain mediating zones between the domains of bush and saltwater groups.

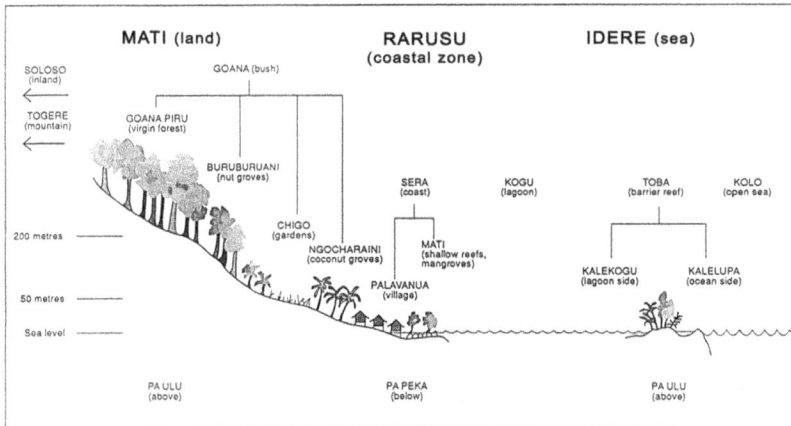

Figure 2.3 Cross-section diagram of the Marovo *puava*.
Source: University of Bergen.

Let me now expand the interpretive dimension of the concept of *puava* in order to note that it is fundamentally the word for 'earth', but, in its wider territorial reference frame, applies to all that which is owned by a *butubutu* as ancestral title, including land, rivers, reefs, lagoon islands and sea (see Figure 2.3). Some *butubutu* of the coast or of salt water, then, have mainly lagoon and reefs as their holdings, whereas other *butubutu* of the bush control large tracts of land on the high islands. According to a foundational environmental classification scheme that applies to the domain of the wild (*piru*), and takes priority over the more domesticated *puava* concept that has the inhabited seashore as its central focus, Marovo people conceptualise and speak of the distinction between *mati* (dry land) and *idere* (sea, salt water), with *rarusu* (coast) as an intermediate zone subject to tidal influences. At their outer perimeters, both *mati* and *idere* are wild, in that interior lands and watercourses, as well as outer reefs and the ocean, are beyond the central domesticated (*manavasa*) zones of the *puava* that cover the lower cultivated foothills, the inhabited or cultivated coast, and the lagoon. The domesticated zones are located 'below' (*pa peka*) in relation to the wild zones 'above' (*pa ulu*). In this sense, rivers and fresh water (*kavo*) are integral to the overall zone of *mati* and, in their spatial extent, cross-cut the topographies of the wild and the domesticated. On a more specific level, rivers form a distinct water domain from that of the sea. The distinctiveness of this opposition, as well as the cultural divide between people of the sea and of the bush, remains today, despite universal coastal settlement.

River Stories, Past and Present

The large mountainous islands of the New Georgia group have rugged volcanic landscapes that include extinct craters with towering rims, deep valleys, pinnacles of raised limestone and tall peaks, rising in places to over 1,000 m. Clouds characteristically hover in the New Georgia mountains and, while weather at the coast and in foothills has a microclimate with significant seasonal variation, rainfall at higher altitudes is much more regular. According to estimates provided by the Solomon Islands' meteorological services, annual rainfall in places such as the crater summit of central Vangunu, where elevations exceed 1,000 m, may approach 9,000 mm. Massive interior rainfall feeds the headwaters of a great number of rivers in the New Georgia and Vangunu mountains. Through stony shallow rapids, slow-flowing meanders and countless streams and small tributaries, water is channelled by the volcanic topography into the major rivers that wind their way through deep valleys and low-lying swamp basins towards expansive mangrove-fringed estuaries, thereby providing the ecological and cosmological connection between the land and the sea. Since all is subsumed in the *puava* concept, the rivers, in an important sense, cross-cut the generic *puava*, provide convenient and unambiguous political boundaries between specific named (and owned) *puava*, and have a reach of environmental influence across the lagoon all the way to the barrier reefs and, ultimately, into the open ocean.

Since my specific geographic focus is the eastern part of New Georgia referred to as the Marovo Lagoon, a closer explanation of the area's quite unique topography is needed. The vast reef-and-lagoon sea spaces of Marovo cover around 700 km². A raised barrier reef extends for about 100 km from the northwestern to the southeastern entrances to the lagoon, running parallel to New Georgia and Vangunu and terminating at the northern tip of the high island of Gatokae. The Marovo barrier reef takes the form mostly of long, narrow and twisted raised islands covered in a particular botanical configuration of lowland rainforest in many places fringed by mangroves. Two-thirds of the barrier reef that forms the Marovo Lagoon follow the northern coast of New Georgia, mostly at distances between 1 and 2 km from the mainland coast; however, at the southeastern end of New Georgia, the width of the lagoon increases to more than 8 km, ultimately forming a globally unique double barrier reef along the lagoon's eastern perimeter. A subsidiary raised barrier reef extends between Gatokae and the southeastern tip of Vangunu, almost enclosing the vast lagoon.

Throughout its extent from north New Georgia to Gatokae, the raised barrier reef is intersected by more than 20 open passages (*sangava*) from lagoon to ocean. Most are fairly deep and navigable, but some are shallow with sandy reef flats and small islets. The passages by the ocean provide a particular connection to the volcanic topography at the other side of the lagoon. In fact, a closer look at a map (e.g. Figure 2.2) shows that all major deep passages are spatially aligned with the estuaries of major rivers on the high volcanic islands. This reflects a long-term environmental process whereby freshwater influences from the rivers have inhibited the formation of coral in those areas where river currents flow towards and meet the barrier reef. Quite simply, the barrier reef passages have been made by the rivers. Several folk tales of New Georgia metaphorically invoke the forces of rivers in the creation of passages through the barrier reef in the form of powerful crocodiles, legendary beings of particular named rivers that have had reason to cut through barrier islands to create openings to the ocean (e.g. Hviding 1995: 74–80). Another important social and cultural manifestation of the topographical singularity of river-passage correspondence is that marine boundaries between *puava* are conceptualised as invisible lines extending across the lagoon from mainland rivers to the adjacent barrier reef passages.

Nevertheless, while great stories are told and complex narratives can be constructed about the land and sea of Marovo, the rivers have received less attention, whether in terms of oral traditions, academic studies or international conservation efforts—the latter having been a major scene of contestation in the Marovo Lagoon since the 1990s (Hviding 2003b, 2006). This lack of attention belies the long and complex cultural histories of rivers, highlighted by the past role of their upper reaches as water sources for large-scale irrigated taro cultivation in pond fields called *ruta* (see Bayliss-Smith and Hviding 2012, 2015), by traditions that tell of bush people with no access to canoes using the river banks as fast access routes between inland settlements and coastal locations, and by recent conflict between *butubutu* of bush and saltwater orientation over logging and associated river-carried sediments. The latter is a particularly contentious present issue.

There are indications that the geological characteristics and scale of ecological systems in New Georgia have always allowed large rivers to carry sediments into the sea. For example, in 1893, Commander A.F. Balfour of the Royal Navy, whose ship the HMS *Penguin* was engaged in an extensive hydrographic survey of the New Georgia group, examined the wide, almost landlocked bay known as Viru Harbour on New Georgia's south coast. Having taken the large ship in by steam through the narrow entrance and anchored in deep water, he noted that the bottom of the bay was covered in black mud, surmising that there had to be 'a river somewhere as the surface water is fresh and very much discoloured. The anchorage is very small but good the bottom is coral with a deposit of [b]lack mud on the top …' (Balfour n.d., entry for 3–4 November 1893).

With a narrow passage to the ocean, Viru Harbour is obviously vulnerable to sedimentation, and the black mud observed from the *Penguin* had been deposited in the deep bay by no less than three major rivers emerging from the volcanic hills. This strong terrestrial connection of the deep landlocked embayment has been perceived as the main reason for siltation during a long period of logging operations from the 1960s onwards, but, as evident from Balfour's notes, the harbour already had a mud-covered bottom and discoloured water in 1893.

While river-carried sediments may always have been a feature of the lagoon environments around New Georgia, the scale of deposits and density of silting has increased in many areas, as logging operations since the 1980s have created deforestation and soil erosion in most of the foothills of New Georgia, Vangunu and Gatokae (Hviding and Bayliss-Smith 2000; see Figure 2.4). The most recent impact of this post-colonial environmental entanglement was the sudden large-scale death in 2011 of marine organisms (from molluscs to dolphins) over much of the central part of Marovo Lagoon, caused by an algal bloom from heavily silted inshore areas (Albert et al. 2012). This disaster caused angry reactions from groups of saltwater people who live on small islands and have customary tenure over most of the lagoon and barrier reef, but who own little land and, thus, unlike the bush people (who receive timber royalties from logging companies), reap few financial benefits from logging. The twenty-first–century relationship between land and sea, now emerging as a matter of political ecology in terms of the river carrying destructive effluent into the sea, is set to intensify throughout Marovo, and to generate more acrimonious conflict between land-holding and sea-holding kinship groups as logging operations continue to expand. Figure 2.4 shows impacts on the land and lagoon of logging operations in the hills, with river-carried sedimentation from the estuary at the lower left of the image. Across the lagoon, the forested barrier reef islands are visible in the upper right-hand corner.

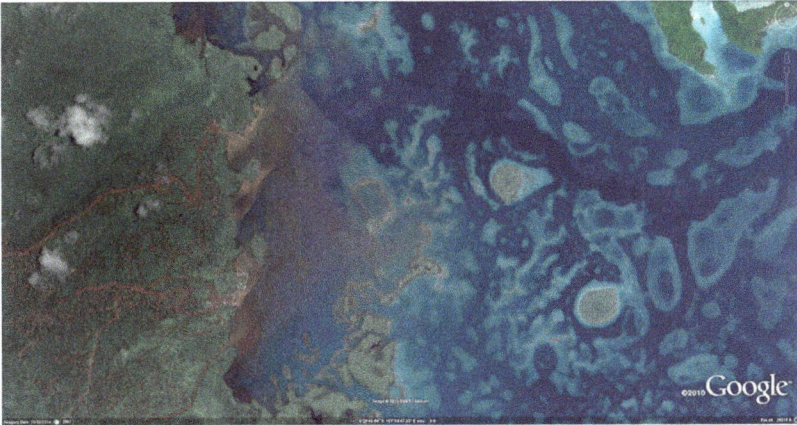

Figure 2.4 Satellite image of the Kolo River area in the western central part of Marovo Lagoon.
Source: Google Earth, October 2014.

Some historical glimpses of rivers as scenes and contexts of human practice are in order here. For the bush people of old Marovo, the lower reaches of major rivers were fishing grounds where hand spears and bow and arrow were the chosen technologies. According to traditions, the bush people rarely possessed canoes, and so no fast water transport between inland and coast was possible for them. However, the banks of streams and rivers provided paths along which bush people could move to the coast and back, sometimes to carry out food exchanges with the saltwater people, who, outside of regular barter occasions, were their enemies and likely to procure slaves and heads from the inland. The act of walking down from hillside settlements, following routes along streams and rivers and arriving at the seashore, is referred to in Marovo as *horevura* ('descend-and-emerge'). This term is also used to refer to the ultimate mass migrations that took place between 1890 and 1920, bringing bush people down to the seashore for permanent settlement once headhunting and intra-Marovo warfare had ceased (Hviding 1996: 97–9).

Long before this process of permanent *horevura*, however, men from the bush would use streams, rivers and tidal creeks and swamps as scenes for productive activity to supply their hillside households with fish, molluscs and crustaceans—much-desired food that is not available in the forest. The main rivers and their adjacent watercourses and flooded lands were used to good effect, providing not only fish and shellfish, but also sheltering the fishermen from the view of the saltwater people, as illustrated by this tale from the hills of Bareke in central Marovo:

> There was a family of a mother and five children. The four brothers made bows. They made arrows to shoot fish. One day, those four went down through the bush from the mountain above Seke in order to descend and emerge near the point at Seke. As those four brothers walked along the Eleluku stream, they continued fishing until they came out at the mouth of the big river Kolo. Then they carried strings with fish back uphill, to their mother and sister who stayed behind in the village. (Adapted from Hviding 1995: 49)

Wisely, the four brothers turn back the moment they reach the open landscape of a major estuary, as to 'come out' in this context means 'to emerge into view'.

Lieutenant H.B.T. Somerville of the aforementioned HMS *Penguin* spent several months ashore in coastal villages of Marovo during 1893–95, using these locations as a base from which to survey the lagoon with a small steamboat and a handful of seamen. Somerville did more than hydrography and chart-making, however. He had a copy of the Royal Anthropological Institute's 'field manual', *Notes and Queries on Anthropology* (Garson and Read 1892), and, like quite a few other gentleman travellers of the day, he engaged in some solid proto-anthropology, ultimately publishing his findings in a 60-page article in the *Journal of the Royal Anthropological Institute* (1897). In his account of the relationships between the lives and political domains of the inhabitants of the Marovo Lagoon, he noted the following about the local people, who at the time still had a particular reputation for 'savagery':

> Their general demeanour is by most white people said to be 'ferocious', and certainly they are inveterate head-hunters. Our officers, however, never experienced anything but civility, good temper, and occasionally kindness at their hands. The result of their custom of head-hunting has been to drive a certain proportion of 'salt water' folk back into the interior, where the tropical density of the bush, and maze of tracks, ensure their safety. I have no doubt that this habit, continued from time immemorial,

has given rise to an opinion (derived from contempt of a foe who hides, rather than fight for his head) that 'man-bush' belongs to a different, and insignificant race. One short excursion that I made into the interior apprised me of the fact that part of the [New Georgia] group, anyway, instead of being very sparsely populated in a few villages on the coast, as generally supposed, is, on the contrary, fairly well inhabited in the interior slopes and valleys of the hills where, in quite a small radius, huts and clearings appeared on all sides in the midst of the bush; quite invisible, however, to a passing ship or canoe. (Somerville 1897: 358)

This, then, was the situation at hand before the process of *horevura* began, and, as already indicated, the divisions remain—although all Marovo villages are now on the coast. I have spent most of my three-and-a-half years of fieldwork in Marovo living with the descendants of Somerville's 'inveterate head-hunters'. This has given me a special vantage point from which to explore the rivers of the rugged mountainous mainland that forms the dark, often cloud-covered backdrop to the beach locations of the saltwater people. Quite simply, saltwater people tend to dislike rivers. They perceive, and treat, mainland rivers as a spatial domain distinctively opposed to the sea and, to some degree, a hostile realm to them. Predictably, and from my subjective point of view as an anthropologist in Marovo *raised* by them as a saltwater man, after 30 years of regular visits I still have not come fully to terms with the dark, muddy domain of the large rivers of New Georgia and the strange and dangerous creatures there. I still find those rivers to be fascinating, exotic and somewhat frightening worlds. This has only partly been remedied by the many shorter stays I have had in the bush village of Tamaneke in northern Marovo, which is itself located in an estuary.

While everyday life and travel on the lagoon coast and out at sea is dominated by a wide open sky, endlessly bright and sometimes harsh and glaring sunshine, and cooling (sometimes even cold) gusty winds, human movement in the domain of rivers is dominated by the dark verdant galleries of estuarine mangroves and riverbank forests, a partly obscured sky and often bursts of torrential rain, particularly in the upper reaches of rivers closer to the inner lands. From the bright and windy world of the sea, one enters the shady and calm world of the river, which may at times even provide sheltered refuge for sea travellers caught in a squall. Even saltwater men dedicated to maritime practice at times admit that the coolness of a large river can provide a pleasant alternative to the scorching heat on the lagoon and the ocean. While rivers and fresh water are cold (*manobu*), the sea and salt water are warm (*reka*). While coldness is not generally desired, it can be so for situational comfort or, most notably, for medical purposes, since some rivers of Marovo have a reputation for having water with strong healing powers in their upper, shallow and faster-flowing reaches. Even reluctant saltwater people may seek out such places when disease necessitates it, courtesy of the knowledge and guidance of their bush neighbours or in-laws.

Piongo Lavata: Travelling the Great River

In the Kalikolo district in the northern reaches of the Marovo Lagoon, canoes powered by outboard motors can reach extraordinarily far inland along major rivers such as the mighty Piongo Lavata.[1] The huge water system of Piongo Lavata meanders up from the coast through a great basin of flat swampland for almost 15 navigable kilometres, with a great number of tributary creeks, streams and swamps intersecting the tall galleries of riverside vegetation. Throughout my fieldwork, I have lived for shorter periods in the central Kalikolo village of Tamaneke, and the bush people of the Vahole *butubutu*, who have been my hosts there, have humorously and generously taught me, in my stereotypical capacity as a saltwater man with little fondness for rivers, how to travel in, and even to some degree appreciate, the dark, muddy and imposing world of their big waterways. The Vahole people number only about 400, but their *puava* is one of the largest in all of New Georgia and, from barrier reef to mountain peaks, covers an area of more than 200 km². The Piongo Lavata is the heartland of their huge *puava*, and their ancestral history is enshrined in the river and its surrounding forests.

Behind the mangrove maze and changeable mud flats of the estuary, the Piongo Lavata's muddy banks are obscured by dense stands of mangroves in the wide tidal zone and of nipa palm just further up. The brackish swamps that extend here on both sides are a zone for extensive food gathering, mainly by women, and offer huge stocks of mud crab and *Polymesoda* and *Anadara* bivalves. In the true freshwater zone that follows, tidal influences are less significant. The distinctive hyper-organic smells of mangroves give way to the earthy sensory impressions of true river mud and of fragrant riverside leaves and flowers. The river banks become more

---

1   The name means 'Great River' in the Hoava language spoken in Kalikolo.

exposed, and here and there one finds easy access to locations known for the presence of large freshwater eels in underwater burrows along the riverbank. It is in the lower reaches of the freshwater zone, away from the mangroves where the forest galleries are at their tallest, that substantial clearings have been made in recent years, both by logging operations (mainly preoccupied with stands of the huge freshwater swamp tree *Terminalia brassii*) and by a proliferation of new food gardens along the river banks where the soil is extraordinarily fertile (Figure 2.5). These garden lands are, on at least an annual basis, flooded by water carrying nutritious sediments. Some say that with the intensification of logging in the foothills, in many cases very close to river tributaries, the flooding of the Piongo Lavata has become more frequent.

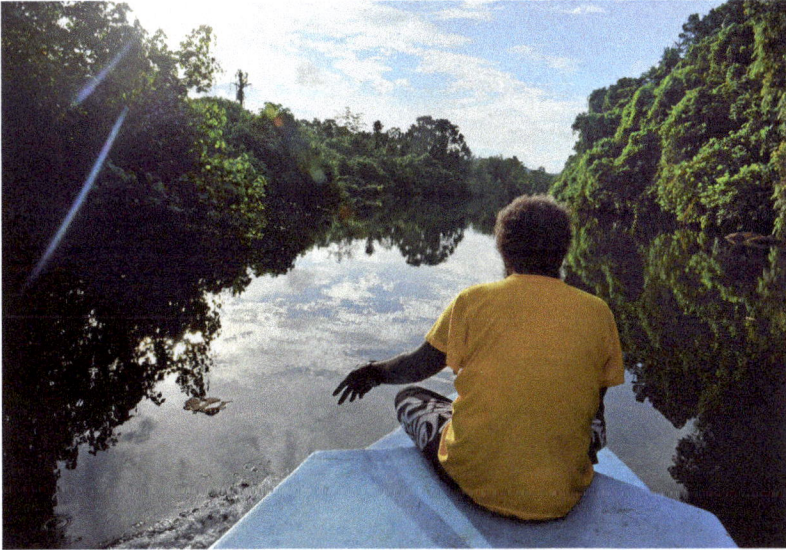

Figure 2.5 Navigating the lower middle reaches of the Piongo Lavata River.
Source: Photo by the author, 2012.

Further up (although this may still be not much above sea level, since most of New Georgia's major rivers only rise quickly through rapids quite far inland), the river traveller is enveloped by dark verdant forest, which, as one gets higher, increasingly forms a rather closed canopy and shuts out the light (Figure 2.6). Noises made by great flocks of parrots reverberate in this confined space, and any sound, such as the movement of a paddle in the water and along the side of a canoe, seems to be amplified and echoed. But this is still a domain of deep cultural history. Even the largely intact riverside forest further inland contains, for the initiated eye, many signs of the dense settlement and cultivation of these extensive lowlands in the past. Groves of *Areca* (betel nut) palms, or tell-tale secondary forest trees and shrubs, which at both canopy and ground levels provide evidence of past vegetation clearance, all point to the density of interior settlement in the past. In this remote part of Marovo, people lived inland for much longer than in the central lagoon area. The Vahole people were masters of *ruta* (irrigated terraced taro cultivation) and, into the twenty-first century, some small irrigated taro plots were still being cultivated along tributary creeks of the Piongo Lavata. Today, the special pond field types of *Colocasia* taro, called *talo ruta*, are still present in the middle freshwater reaches of the Piongo Lavata. Some have been planted secretively for safekeeping of the species, but *talo ruta* increasingly betrays this purpose by going feral and turning up in riverside locations, reminding observant passers-by of the history of inland settlement and cultivation (Figure 2.7). The criss-crossing of the riverside landscapes by signs of past habitation and cultivation, often in the form of sites associated with named ancestral persons, is a powerful demonstration of customary entitlements to the Vahole land.

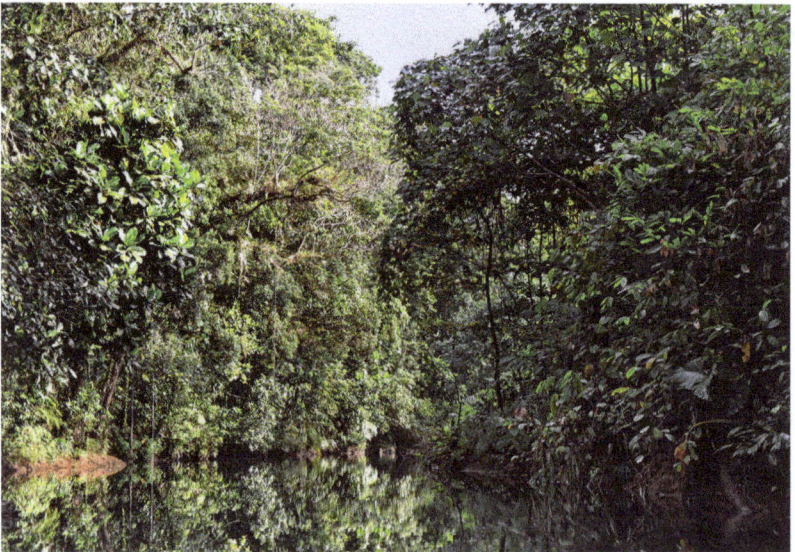

Figure 2.6 Largely undisturbed river forest in the upper middle reaches of the Piongo Lavata River.
Source: Photo by the author, 2012.

Figure 2.7 Wetland pond field variety of *Colocasia* taro in feral state.
Source: Photo by the author, 2012.

This far up the river, broken branches stuck on the bottom and volcanic boulders in the water are hurdles to navigation, but, as the water gets less turbid, its clarity allows for an easier view of such underwater obstructions. Given today's proliferation of crocodiles, this zone of clear water flowing with increasing speed over a sandy riverbed (Figure 2.8) is the only place in the river where people will relax their continuous alert for the predators. Crocodiles have, in recent years, made quite regular attacks on women working in their riverside gardens in the lower middle reaches of the river. Those who have survived have, in most cases, lost an arm, since the crocodile tends to attack women as they wash the day's harvest of root crops by the river bank. In the upper middle reaches, crocodiles are not infrequently seen resting on logs and mud flats along the river banks. But nearer to the rapids it is a different matter. 'Up here, you can see a crocodile as easily as it sees you', the saying goes, as far as the end of navigable water is concerned. This observation is helpful, since it is this cool, crystal clear and more rapidly flowing water in the upper reaches of the Piongo Lavata that is considered to have healing properties. Whether you are healthy or not, a swim is always recommended on the rare occasion when visits are made to this remote zone.

Figure 2.8 In the upper reaches of the Piongo Lavata River.
Source: Photo by the author, 2012.

Not far from the location shown in Figure 2.8, the Piongo Lavata narrows and rises quickly. Shallow water runs in rapids over boulders and, from this spot, water transport can only be obtained by first carrying small dugout canoes up through the rapids. However, 15 km from the coast as the pigeon flies, beyond the stony rapids and at an elevation of about 150 m, the Piongo Lavata's headwaters are still not near. Fed by a multitude of tributaries that cascade down from the tall ridges of an old crater, the river meanders slowly for several kilometres, across a relatively flat upland basin where the water is fringed and partly hidden by dense tall forest not yet touched by logging machinery. Here, the Piongo Lavata forms wide, quite deep expanses of slow-moving water partly covered by the forest canopy. These are referred to as lakes (*kopi*, a term also applied to substantial reef basins out by the ocean), and have shorelines of volcanic boulders. No logging has taken place up here; the forest and the rivers are in pristine condition, and people very rarely go there. In fact, only a few men are likely to reach this far, in the course of hunting the feral pigs that are abundant near rivers in the upland forest—most people of New Georgia will never move this far inland. This remote place is also considered to be the home of a very special large animal, a somewhat mystical crocodile of the inner lands, considered locally to be a separate species. Past the spectacular upland river lakes are the proper headwaters of the Piongo Lavata. The great river has its beginnings on steep mountain ridges rising up to 400 m above sea level, with a singular vegetation of particular *Pandanus*, *Casuarina* and ginger species and mosses hanging from the branches of stunted trees, and with a specific repertoire of song from mountain birds.

The headwaters of most rivers of New Georgia emerge in such high-elevation places from clusters of smooth volcanic boulders, from which small streams meet to form tributary rivers that ultimately converge in one major watercourse in valley bottoms. There is a distinct vocabulary concerning high-altitude headwaters, such as for describing the particular sounds made by clear water that flows forth and trickles over boulders and makes a river begin. Standing on a reasonably clear day among boulders and stunted high-altitude vegetation by the headwaters of a major river like the Piongo Lavata, at the end of what feels like a journey in both time and space, the traveller is likely to be amazed by the sweeping views over huge tracts of hill forest towards the remote lagoon and very distant barrier islands. In central New Georgia, such views can even cover both the Roviana and Marovo lagoons and, in clear weather, the other islands of the New Georgia archipelago are clearly seen, as well as the outlines on the northern horizon of the islands of Choiseul and Isabel. Following a big river to its beginnings demonstrates the scale of place in New Georgia; even for a Vahole person, who may only be up here once in a lifetime, the realisation of the extent of the ancestral land and of the scale of the landscape is particularly acute—the moment is an emotional one. For saltwater people, this is as far as one can get from the ordinary everyday world.

Vua: Chief of the River

The observable lives of crocodiles (*vua*) in Marovo converge in several ways with the life of the river, as exemplified by the Piongo Lavata. The crocodile's prominence in Marovo cosmology and cultural history makes it the 'chief of the river' (*bangara pa kavo*), an adversary to so many human projects, which evades people by stealth until it strikes by ambush. While the bush people of old Marovo had the river as a primary focus of movement and food production, and their present-day descendants maintain an overall orientation to the land and often travel on and otherwise use its rivers, I have indicated how the freshwater environment is viewed by saltwater people with ambivalence, as an environment quite hostile to their preferred way of life. For them, rivers harbour dangerous creatures, most significantly the crocodile, which has rivers and estuaries as its main haunts. In the enduring cultural structures of New Georgia totemism that I have already briefly mentioned, the crocodile is considered to be ancestrally unrelated to saltwater people, who, for their part, consider the shark (in its previously described generic sense) to be their ancestral relation. From an outsider perspective, it would seem remarkable that an animal usually known to the world as the saltwater crocodile would be totemically associated with bush people, but such are the spatial and cosmological distinctions in the world of Marovo. The crocodile is first and foremost a creature of the river, and of the mangrove forests in estuaries and along mainland shores. While it may well be encountered at sea, at times even prowling the barrier reef slopes facing the ocean, maritime places are not regarded in Marovo as its proper habitat.

Fundamentally, then, crocodiles are associated with those groups of bush people who have a strong historical connection to rivers and estuaries, and sharks are associated with saltwater groups whose histories are primarily maritime. The structural symbolic scheme is that in which saltwater people are to bush people as the shark is to the crocodile.

As living beings with ancestral qualities attributed to them, sharks and crocodiles share certain attributes. Those who hold sharks or crocodiles as their ancestral spirits must not insult, injure, kill or eat them. They must not refer to a shark or crocodile as such when sighted, reflecting a pan-Marovo desire to distance these dangerous animals from humans, who must maintain a respectful relationship—including the naming taboo—to safeguard against attack. Even today, any crocodile or shark sighted is referred to as 'something', its whereabouts indicated not with hands, but by a directional flicker of the eyes. A crocodile may be alternatively referred to as 'old man' (*maroke*). While crocodile-related bush people commonly claim that such properly respectful behaviour protects them against attacks from the animal, and saltwater people claim that similar attitudes and relationships to sharks protect them against shark attacks, such safeguards are less effective when the roles are reversed. Most documented shark attacks in Marovo over the years have in fact involved people from bush *butubutu*, who also dive and fish with spears, and thereby enter the realm of the saltwater people, but do not possess their totemic protection. Saltwater people, who travel and fish in places that used to be largely free of crocodiles, are increasingly experiencing threats from them. Today, crocodiles seek out village shores, where dogs are easy prey at night, and spear fishermen from saltwater villages have, on occasion, even encountered crocodiles along the ocean-facing reef walls. It is said that crocodiles are no longer afraid of humans, since no one now hunts them, and nothing inhibits the *vua* from ranging far outside its river domains.

While an assessment from surveys carried out in 1989, based on local knowledge and information from crocodile hunters, concluded that 'only a widely scattered remnant number of crocodiles remain [in the New Georgia Islands]' (Messel and King 1990: 47), this is no longer the case. During my fieldwork in Marovo Lagoon in 2010 and 2012, people commented that crocodile populations had bounced back rapidly and were now at the level of pre-colonial times, when there was no hunting of the reptiles and they were held in sensible reverence even by saltwater people, owing not least to the proverbial habit of large ones to float threateningly close to village shores at dusk (*irongo pa veluvelu*). An old man commented to me: 'Crocodiles were abundant here in the old days. But now they are just as numerous again, and no longer afraid of humans!'

It is reasonable to assume that saltwater crocodiles, whose diets are diverse and well covered by the abundance of animal life in the rivers, mangroves and reefs of the larger Pacific islands, were numerous and truly menacing for the early settlers in the Melanesian archipelagos and further afield. The reptiles also had a greater Pacific range in the past: a review of extensive pan-Pacific materials by Best (1988) points to the enduring impact of Pacific Islanders' historical relationships with crocodiles, exemplified by Māori and other Polynesian traditions of giant man-eating lizards and similar mythical creatures, which may constitute society's long-term memory of human–crocodile relations.[2]

In the late colonial era, crocodiles were extensively hunted and decimated throughout most of the Solomons and so came to represent less of a threat. When I first came to live in Marovo in 1986, crocodiles were encountered infrequently and were not regarded as much of a danger, as long as one kept away from places known to be their habitual hideaways. The local hunting of saltwater crocodiles in estuaries and swamps at night, with rifles, spears, torches and lamps, and sometimes with the assistance of dogs, had been an enduring, regular activity around New Georgia. Pursued by those brave or foolhardy enough, and preferably

---

2   See also Torgersen (this volume) on the lethal *mo'o* of the Wailuku River in Hawai'i.

without ancestral relations to the crocodile, this had its origins in visits by Australian hunters during the 1950s. Surprisingly, there are no accounts in Marovo of a local crocodile hunter ever being killed or injured by his prey, even if the dogs were not so lucky. The saltwater crocodile has the most valuable of all crocodile skins, and, from the 1960s into the 1980s, local hunters could make substantial money from the sale of skins to entrepreneurial middlemen. The end of hunting came in the late 1980s when the implementation of global crocodile conservation measures cut short demand, and no skins seem to have been traded in the Solomons since around 1990, except for those obtained from a small number of farmed animals. Whereas the aforementioned surveys carried out at the time concluded that the crocodile population was verging on extinction, assessments by villagers were not that pessimistic and, while the ban on the export of crocodile skins has remained in force, local populations have been booming in many parts of the country.

Even the occasional defensive shooting of crocodiles taking up residence too close to villages stopped altogether from 2003. That year saw villagers throughout the Solomons surrender all firearms through large-scale national disarmament by the Australian-led Regional Assistance Mission to Solomon Islands (RAMSI), which entered the scene after armed conflict had ravaged parts of the country since 1998 (Moore 2004). No longer are shots fired at crocodiles, except in those rare and bizarre cases when armed RAMSI soldiers have been dispatched by air to the scene of a crocodile attack, where they are usually unable to find the guilty animal or, if one is indeed located, to actually hit it with their high-tech automatic weapons.

In New Georgia, women, whether paddling on the lagoon or working in riverside locations (such as those already described), and children, paddling to school or playing by the shore, have been victims of the majority of attacks from the growing crocodile population. Crocodiles seem adept at capsizing a small dugout canoe and grabbing the hapless paddler; it is believed that they are clever planners, able to observe patterns of regular travel and to devise an appropriate ambush strategy. Even some men engaged in underwater spear fishing at the barrier reef have been attacked, although these are considered unlucky chance encounters caused by increased crocodile mobility. Some victims have disappeared without a trace, the mangled, decomposed remains of a few have been retrieved, and some have escaped alive but with serious injury caused by the crocodile's 'death roll', a violent movement aiming at ripping large prey apart.

Quite simply, people say that the crocodiles of Marovo have become fearless and ruthless. The predators have really honed the skills of stealth and ambush that they were prevented from developing so long as people shot at them and made them hide in the mangroves. Today, the *vua* is everywhere—in the upper and lower reaches of rivers, in mangroves, on village shores, in the lagoon and around the lagoon islands, and even in the barrier reef. Anyone who goes to the seaside at night for toilet needs (and nearly everyone still has to do so) must now bring a torch and watch out for whatever is lurking near the shore or under overhanging trees. In some villages, where children used to paddle to school in the morning in small dugout canoes, group transport by motorised boat is now the cautious strategy. On another level, it is increasingly claimed today that the crocodile no longer follows its prescribed behaviour according to the totemic contract with those *butubutu* with which it is associated, since even people from those groups are now being attacked. Some say the crocodile spirits are angered by the environmental destruction caused by logging. Some survivors of crocodile attacks belonging to crocodile-associated groups have told tales of a human-like spirit being hovering as an opaque shadow nearby while they were in the animal's jaws—a reference to the growing belief that sorcerers also take advantage of the present abundance of crocodiles.

'It is we humans who are now the endangered species', is a comment heard throughout Marovo in the twenty-first century, with a wry reference to the local human consequences of global conservation. Similar observations are made throughout the Solomons in places where rapid increases in crocodile attacks are experienced. On 26 July 2011, the news site *Solomon Times Online* carried this frank assessment from the government:

Crocodile Population a Worry: Ramofafia

The Solomon Islands Ministry of Fisheries says it will call an urgent meeting with the Ministry of Environment to discuss the country's growing crocodile population.

The crocodiles are an increasing threat to livelihoods and in some cases lives. The numbers of crocodiles in Solomon Islands started to increase after the trade in crocodile skin was banned in the early 1980s. The situation has become more challenging after the country became gun-free with the deployment of the Regional Assistance Mission in 2003. Dr Christian Ramofafia, Permanent Secretary of the Ministry of Fisheries and Marine Resources, acknowledges the problem and says they are doing all they can to address the problem. An accurate count of annual crocodile attacks on humans in the Solomon Islands is somewhat difficult to obtain. Many of the areas in which humans and large crocodiles come into contact are in remote parts of the Solomon Islands. Crocodile attacks are therefore not always reported to local authorities, and some reports are difficult to verify. (Anon. 2011)

Flows, Directions and Inhabitants

Despite the present overflow in terms of population and territory, the saltwater crocodile remains an icon first and foremost of New Georgia's rivers, being the most prominent and most dangerous inhabitant of this particular environmental domain. But there are also other creatures that approach such iconic roles, though not representing a danger to humans. The large freshwater eels are a case in point; their attractiveness as a particularly 'greasy' protein food and the singular methods of locating and catching them remain firmly in the domain of bush people. The mullet, on the other hand, which forms huge seasonal schools in the lower reaches of major rivers, is iconic for the river activities of saltwater people, who are the most adept in finding and netting those schools. While the mullet is viewed as a fish of the sea and a sought-after source of prestigious food, this is not so for the plethora of mainly small, truly freshwater fish. While some bush people are known for catching even the smallest freshwater fish for everyday food, most of them, and all saltwater people, consider freshwater fish to have a weak and watery taste, and to be food only for the elderly and sick or for pregnant women. Mangrove snapper, the only river fish of some size, is widely disliked and avoided as food because of its habit of eating snakes.

There are also other fishes known in areas of New Georgia to be characteristic inhabitants of certain waterways or of rivers in general. Some pointed out to me during fieldwork do not readily conform to existing scientific overviews of the freshwater fauna of the Solomon Islands.[3] In general, the rivers of the Solomons are poorly studied by biologists, and new species are often found when surveys are done. After a rewarding trip to the island of Choiseul, across the sea from New Georgia, David Boseto (the only Solomon Islander freshwater biologist) and his research

---

3   A type of colourful smaller eel with a laterally flattened body and pointed snout, swimming in the clear middle reaches of Piongo Lavata, comes to mind.

team concluded that '[t]he low freshwater fish diversity recorded for Pacific Island countries can, in some cases, be attributed to the absence of research being carried out' (Boseto et al. 2007: 16).

The rivers of New Georgia, then, are scenes for intense interaction among differentiated social groups and ecological domains and the living creatures there, which may be more or less well known to science. The river is always in motion, but so is the sea; every lagoon traveller takes into account a complex knowledge of currents and tides. But the river is unique in that it connects the land with the sea, differing from the land/sea connections made at the seashore and in the tidal zone. When there is high tide, the large rivers of New Georgia may have a surface layer of salt water several kilometres inland; conversely, at low tide, the brackish estuaries may be largely fresh some distance out into the lagoon along the mud flats. The river is seen by Marovo people as a unidirectional carrier of water, creatures, sediments and anything else from its terrestrial domain out into the sea, thus bringing the substance of *mati* into *idere* (see Figure 2.3). However, that is but one direction of the river's flow. The other is associated with human agency and involves travel from the coast towards inner lands, actively against the river's current (although a wise paddler in a small dugout canoe is not likely to try to reach far upriver when the tide is going out in the estuary). The constantly changing ebb and flow of the river, the massive stature of the riverside galleries of mangroves, nipa palm and tall rainforest, and the movement, both downriver and upriver, of non-human creatures and substances and human beings themselves all make for a world strikingly different from the sun-drenched openness of the lagoon seascape. In a cosmological scheme where that which is dark and cold is contrasted with conditions of light and warmth, navigating up one of New Georgia's rivers becomes a journey into the unpredictable and potentially antagonistic, especially for saltwater people. But even people whose historical connections are inland—of the bush—may take such views, considering the current universality of coastal village life on sunny lagoon shores.

I shall round off my account with a discussion of freshwater crocodiles, which, according to the received wisdom of biology, do not exist in the Solomon Islands. Biologists instead believe that the smaller crocodiles living in the upper freshwater reaches of rivers are simply young saltwater ones that seek refuge from voracious larger relatives. As noted by the foremost authority on the reptiles of the Solomons:

> Some popular travel guidebooks on the Solomons mention a second 'harmless' species of freshwater crocodile occurring in rivers. This is an incorrect and potentially dangerous piece of misinformation. There is only one species of crocodile in the Solomons, the saltwater crocodile, and even small specimens are far from harmless. (McCoy 2006: 16)

Nevertheless, in the vernacular conceptualisation of crocodiles in Marovo, two distinct species exist, and the notion of harmlessness is equivocal as far as the freshwater crocodile goes. I have noted that, while the saltwater crocodile or *vua hokara* ('proper crocodile') belongs to the realm of the bush people and is the close historical associate of many *butubutu*, it also leads a coastal existence and figures strongly in the traditions of a number of saltwater groups, who have no 'contract' with the *vua* that protects them from attacks. But saltwater people believe that another, much less common crocodile of true fresh water is capable of causing worse harm than the normal crocodile.

The usual story goes like this. If you venture far inland, up the major rivers into the upland basins where the water moves slowly in deep, calm meanders, as exemplified by the Piongo Lavata, you enter the territory of the *vua varane* ('warrior crocodile'). While the normal *vua* is known in Marovo to reach formidable sizes of 6 m or more, even while those of about 3–4 m are considered to be the most hungry and dangerous, the *vua varane* is usually only about a fathom long, with only a few approaching 3 m in total length. Unlike its broad-faced, heavy-set saltwater cousin, it has long narrow jaws and a nimble body, and it is faster and more aggressive— hence its warrior label. Tales are told in the traditions of saltwater people about what could happen to raiding parties venturing upriver to take the heads of bush people. The *vua varane* could snatch warriors walking along a river bank and drag them away to be eaten. In an important sense, the *vua varane*—a true freshwater crocodile where, according to biology, there is none—stands for the dangers of the upper river and interior lands, in ways that reverse the time-honoured hierarchy in which saltwater people dominated bush people. If you are a saltwater man, you should only venture so far into a river as not to risk encountering the *vua varane*, which, unlike the bush people, has no fear of you. What, then, do bush people have to say about the *vua varane*? They claim to have no such animal—at least not one that is known by that name. True, they too insist that there are real freshwater crocodiles up there, with longish narrow jaws and nimble bodies, but their name for them is *kapakale*—a term that refers to their habit of resting on a stone or a submerged log. It is not regarded by them as dangerous, but they are well aware the saltwater men fear it. Like the environmentally malevolent transport of sediments into the lagoon, the mysterious freshwater crocodile is the river's triumph over the sea.

Taming the Water: From River to Village

Ideas of power, dominance, domestication and control, as exemplified above by the *vua varane*, are also at the core of some striking present-day reconfigurations. While I have discussed at length the environments and creatures of the river, and the role of the river in a cosmological scheme of relationships between land and sea, and the associated groups of people, in this final section I turn to a more direct engagement with rivers that takes on growing significance in twenty-first–century New Georgia. I move to a rural technological dimension, focusing on how a current efflorescence in New Georgia of quite advanced rural water supply systems is interpreted locally as a way of harnessing the forces of rivers for greater human comfort. And so I end my account with the more mundane aspects of fresh water in everyday village life.

Everyone in Marovo—except those who have access to rainwater tanks connected by gutters to corrugated iron roofs—depends on water from rivers, streams, springs and non-stagnant freshwater pools for drinking, cooking and washing. The high rainfall and massive network of natural watercourses offered by the rugged topography of New Georgia have made for relatively easy access to high-quality drinking water. But actually bringing the water to the houses has been more of a challenge.

Villages on the so-called 'weather coast' of the island of Vangunu are located near to sandy estuaries where large rivers enter the ocean. With the coastal land rising steeply just behind the narrow coastal plain, access to clean fresh water has always been easy in this part of New Georgia. For some coastal lagoon villages with a convenient location close to waterfalls, simple gravity-fed water supply systems, with a few standpipes distributed among the houses, had already been introduced in the 1980s. Stories were told around the lagoon of how those novel standpipes provided cool, clear water from waterfalls at pressures that could make a shower almost painful. This technological development, at first limited to a few fortunate villages, represented pioneering interventions into the river's flow of fresh water, which had until then been perceived by the majority of Marovo villagers as a resource subject to rather complicated access from sources far away from where they lived. Piped river water supplies, still gravity-fed from somewhat higher altitudes, became widespread throughout Marovo during the 1990s, and effectively replaced labour-intensive and inefficient fresh water procurement. No longer was it necessary to walk to water sources, and no longer did entire villages have to depend on precarious springs or stagnant pools for fresh water. After the proliferation of standpipe systems in villages on the mainland shores of the Marovo Lagoon, the only remaining exceptions to the new rule of easy water access were the villages located on small waterless coral islands, where only those households with corrugated iron roofs could rely on rainwater tanks.

This ongoing improvement of village water supply throughout the Marovo area has reconfigured the enduring saltwater/bush distinction in significant ways. Whereas the lack of easily accessible fresh water used to be emblematic of the lives of saltwater people on small coral islands, the recent and present proliferation of innovative approaches to controlling and redirecting the flow of water for village purposes has now made abundant fresh water a fact of everyday life for almost everyone. New arrangements whereby saltwater villages on small islands off the mainland have obtained access to gravity-fed water supplies piped across the lagoon bottom from mainland rivers have, for example, involved significant rearrangements of gendered labour patterns. However, the cultural distinction between salt water and bush remains in place, since the expansion of piped water supplies to offshore islands has invoked new disputes over water ownership in cases where the recipients of the water do not own its source.

A striking example of technological innovation in mainland–offshore connections through fresh water, and consequent transformations of everyday life, comes from the barrier reef village of Keru in northern Marovo. Keru is located on a raised coral island with a jagged limestone topography, and, for many years, only had a seasonally fluctuating community water supply in the form of a rain-fed concrete tank whose contents were rarely fit for drinking. Visits to the mainland for laundry and bathing, mainly by the women of the village, used to involve up to 45 minutes of paddling each way, as well as hauling plastic cans of cool water from clean streams back to the village. Emblematic of the existence of true saltwater people, this had been the situation since Keru was settled around 1920, modified only slightly for a few households that obtained private rain-fed water tanks connected to corrugated iron roofs.

By about 2000, when income from logging operations on the mainland foothills controlled by the Keru people allowed for it,[4] an ambitious plan involving gravity-fed water from the upper reaches of a river was developed. From a small waterfall about a kilometre's distance uphill, a tough flexible pipe was led down to the seashore and made ready for the long reach, through many joined sections, across almost 5 km of lagoon out to Keru. Like Tamaneke, Keru is a village of the Christian Fellowship Church, an Indigenous religious organisation and powerful social movement that plays a significant political and economic role in New Georgia (Hviding 2011), and is also well known for its ability to mobilise massive communal labour. A notable feat of cooperation involving a number of villages of northern New Georgia saw several hundred canoes being paddled, on a fine calm day, into position to form a bridge-like structure between the mainland and Keru. Upon this platform of canoes, sections of concrete tubing were placed, through which the water pipe was led, in the process connecting every single canoe. At a given signal, all canoe crews pushed the concrete-and-pipe assemblage overboard onto the lagoon floor, which in some parts of this area is up to 20 m deep. Participants were somewhat surprised that not a single canoe capsized or sank at that moment. Symbolically, this was a spectacular feat of instantly connecting the river and the barrier reef, the fresh water and the sea, and the mainland territory of the Keru people with their offshore village. It was also about harnessing the water of the river to accomplish a major transformation of village life: no longer were those long hours of travel back and forth across the lagoon to mainland rivers necessary.

The construction of piped water supplies around Marovo, from the simplest to the most complex, is no mundane process in the eyes of its protagonists, but is interpreted as a process of 'domesticating' (*va manavasa*) the initially uncontrolled, 'wild' (*piru*) resource represented by relentlessly flowing river water. Whereas I have explained earlier how the flow of the river cross-cuts the wild and the domesticated, the water itself remains wild. In this sense, taming the wild water also engages another dimension of connectivity: bringing water to places of human habitation where there has not been any previously, thereby harnessing a power of the wild into a productive domesticated force that has the potential for completely transforming relationships of labour and gender in the village. Obviously, bringing river water under control for human purposes today echoes the

---

4   Despite their sea-oriented life, the people of Keru also control substantial forested lands.

old times when the water of rivers and streams was diverted to feed the taro pond fields of the inland valleys. My discussion thus ends here, with an argument that embraces both connectivity and duality in the relationships of rivers and sea, of fresh and salt water—on a contemporary note where new forms of village-level technology are used to once more bring the forces of the river under human control. The river continues to flow—but its force may be converted into novel channels for human benefit.

References

Albert, S., M. Dunbabin, M. Skinner, B. Moore and A. Grinham, 2012. 'Benthic Shift in a Solomon Islands Lagoon: Corals to Cyanobacteria.' In *Proceedings of the 12th International Coral Reef Symposium*. Cairns: Australia.

Anon., 2011. 'Crocodile Population a Worry: Ramofafia.' *Solomon Times Online*, 26 July.

Balfour, A.B.F., n.d. 'Private Journal and Remarks 1893–1894–1895.' Balfour Collection 1873–1896 (ABA/V). London: Royal Geographical Society.

Bayliss-Smith, T. and E. Hviding, 2012. 'Irrigated Taro, Malaria and the Expansion of Chiefdoms: *Ruta* in New Georgia, Solomon Islands.' In M. Spriggs, D. Addison and P. Matthews (eds), *Irrigated Cultivation of Colocasia Esculenta in the Indo-Pacific: Biological, Social and Historical Perspectives*. Osaka: National Museum of Ethnology (Senri Ethnological Studies 78).

——, 2015. 'Landesque Capital as an Alternative to Food Storage in Melanesia: Irrigated Taro Terraces in New Georgia, Solomon Islands.' *Environmental Archaeology* 20: 425–436. doi.org/10.1179/1749631414Y.0000000049

Bayliss-Smith, T., E. Hviding and T.C. Whitmore, 2003. 'Rainforest Composition and Histories of Human Disturbance in Solomon Islands.' *Ambio* 32: 346–352. doi.org/10.1579/0044-7447-32.5.346

Best, S., 1988. 'Here Be Dragons.' *Journal of the Polynesian Society* 97: 239–260.

Boseto, D., C. Morrison, P. Pikacha and T. Pitakia, 2007. 'Biodiversity and Conservation of Freshwater Fishes in Selected Rivers on Choiseul Island, Solomon Islands.' *South Pacific Journal of Natural Science* 3: 16–21. doi.org/10.1071/SP07003

Garson, J.G. and C.H. Read, 1892. *Notes and Queries on Anthropology* (2nd edition). London: The Anthropological Institute.

Hviding, E., 1995. *Vivinei Tuari Pa Ulusaghe: Stories and Legends from Marovo, New Georgia, in Four New Georgian Languages and with English Translations.* Bergen: University of Bergen, Centre for Development Studies, in collaboration with Western Province Division of Culture.

——, 1996. *Guardians of Marovo Lagoon: Practice, Place, and Politics in Maritime Melanesia.* Honolulu: University of Hawai'i Press (Pacific Islands Monograph 14).

——, 2003a. 'Disentangling the *Butubutu* of New Georgia: Cognatic Kinship in Thought and Action.' In I. Hoëm and S. Roalkvam (eds), *Oceanic Socialities and Cultural Forms: Ethnographies of Experience.* New York: Berghahn Books.

——, 2003b. 'Contested Rainforests, NGOs and Projects of Desire in Solomon Islands.' *International Social Science Journal* 178: 439–453.

——, 2005. *Reef and Rainforest: An Environmental Encyclopedia of Marovo Lagoon, Solomon Islands / Kiladi Oro Vivineidi Ria Tingitonga Pa Idere Oro Pa Goana Pa Marovo.* Paris: UNESCO (Knowledges of Nature 1).

——, 2006. 'Knowing and Managing Biodiversity in the Pacific Islands: Challenges of Conservation in the Marovo Lagoon.' *International Social Science Journal* 187: 69–85.

——, 2011. 'Re-Placing the State in the Western Solomon Islands: The Political Rise of the Christian Fellowship Church.' In E. Hviding and K.M. Rio (eds), *Made in Oceania: Social Movements, Cultural Heritage and the State in the Pacific.* Wantage (UK): Sean Kingston Publishing.

Hviding, E. and T. Bayliss-Smith, 2000. *Islands of Rainforest: Agroforestry, Logging and Ecotourism in Solomon Islands.* Aldershot: Ashgate.

McCoy, M., 2006. *Reptiles of the Solomon Islands.* Sofia: Pensoft Publishers.

Messel, H. and F.W. King, 1990. 'The Status of *Crocodylus Porosus* in the Solomon Islands.' In *Crocodiles: Proceedings of the 10th Working Meeting of the IUCNSSC Crocodile Specialist Group.* Gland: IUCN.

Moore, C., 2004. *Happy Isles in Crisis: The Historical Causes for a Failing State in Solomon Islands, 1998–2004.* Canberra: Asia-Pacific Press.

Somerville, H.B.T., 1897. 'Ethnographical Notes in New Georgia, Solomon Islands.' *Journal of the Royal Anthropological Institute* 26: 357–413. doi.org/10.2307/2842009

3. A Source of Power, Disquiet and Biblical Purport: The Jordan River in Santo, Vanuatu

CARLOS MONDRAGÓN

Introduction

The Jordan River, on the island of Espíritu Santo, is the largest river in the Vanuatu archipelago.[1] It was originally given that name 400 years ago by a foreigner, Iberian navigator Pedro Fernández de Quirós, the first of many outlanders who have since engaged with the people and environments that coexist with this body of flowing water. Of course, contemporary Santo Islanders have their own names for this river, as well as for the specific tributaries that give form to its drainage basin. The rugged geography over which this basin extends is located within a broader, geographically complex region of North Santo. This territory, to whose description I return below, is host to a rich diversity of ni-Vanuatu communities, language groups and storied places.[2] Importantly, this is

---

1   For the purposes of this paper, I focus on the part of the Jordan that constitutes the large, slowly flowing body of water that snakes across and gives form to the lowland and coastal delta of Big Bay. This geography includes the 'mouth' of the river, which is characterised by a shifting series of marshes and unstable waterways that empty, with very little strength, into Big Bay itself.

2   The island of Espiritu Santo is home to at least 15, and possibly as many as 28, separate language communities (Tryon 1996; Lynch and Crowley 2001: 43–58).

a territory that is powerfully shaped and characterised by multifarious forms of salt and fresh water; rivers, creeks, waterfalls, marshes, springs and man-made terraces for the cultivation of water taro are only part of the watery phenomena that have produced the North Santo landscape over the course of hundreds and even thousands of years.

The aim of this chapter is to explore the human and environmental diversity of this North Santo waterscape through the layered stories of mobility and belonging by which local communities relate to their watery surroundings. Specifically, my chapter is about water-related geographical identities in relation to the contested past, and to the uncertainties of movement and transformation that are experienced in connection with physical and spiritual sites of fresh water and flowing water. It is worth emphasising that references to these tangible and intangible sources of fresh water—be they the rivers, waterfalls and springs, or watery spirits— are necessarily inflected with morally laden values that sometimes work as powerful, prescriptive reminders of the behaviour that humans and non-humans must exhibit towards each other and towards significant aspects of their circumambient spaces. In this regard, fresh water, especially as personified by the Jordan River and various important springs, is a critical aspect of the mutual constitution of persons and land, understood as a broader, multi-layered phenomenon. Thus, this text is about the genealogical and moral dimensions of the waterscape of North Santo in relation to what Biersack (1990) once defined as the process of 'history in the making', which is premised on land and genealogies of belonging.

Because this chapter is about geographical identity, as opposed to simply 'territorial' identity (Mondragón 2009), it is worth remarking that the exploration of territory and identity has long been the subject of anthropological scrutiny in Vanuatu research (e.g. Lane 1971; Rodman 1987; Bonnemaison 1996; Hess 2009; Mondragón 2012, 2015a, 2015b). These important contributions have given rise to key insights about images of rootedness and mobility among the different island communities that make up the archipelago (see also Bonnemaison 1994, 1997; Jolly 1999; Taylor 2008). Here, however, I foreground water and not land as my primary object of interest. It is my intention to extend understandings of territoriality to the often ignored watery sites of the land by highlighting some of the ways in which these places are enmeshed in the imagination of kinship, human/non-human relations, land use and belonging in North Santo.

My use of the term 'waterscape' in this chapter is partly informed by Baviskar's use of the term in order to evoke layered ecologies, as constituted through historical, environmental and political concerns (Baviskar 2007). But, in contrast to Baviskar's focus on 'struggles over water', my primary interest is to address the themes that recur through this volume by focusing on the neglected moral, spiritual and agentive 'layers' of fresh water, history and cultural landscapes in island Melanesia.[3]

The materials in this paper are intended to begin to address the interleaving of fluid and territorial localities, in time and space, as relevant frames from which to approach the ongoing shifts and transformations that inform North Santo experiences of geographic belonging. In particular, my interest in layered geographies is focused on contested pasts and on the disquiet that accompanies boundary work in the expression of belonging—in other words, the ongoing iteration and contestation connected to places of origin, processes of movement and emerging sites of transformation. The focus on layering is intended as a way of emphasising that human–environmental relations across North Santo are premised on a broad panoply of processes, rather than structures or types. In other words, I am interested in exploring the particular processes that make up the 'living milieu' of North Vanuatu societies (Mondragón 2009).

In the ethnographic and historical materials that follow, I have taken a cluster of powerful toponyms associated with the River Jordan as my point of departure for this exploration of the mutually constituted waterscapes that partially reflect what Fox (2006) has termed 'topogenies', and that inform the watery micro-geography of North Santo.

The North Santo Context

The island of Espiritu Santo, known commonly as Santo, is located in the north-central area of the Vanuatu archipelago. With a surface area of 3,955.5 km², and a north–south length of more than 50 km, Santo is the largest island in Vanuatu (Siméoni 2009). It hosts the archipelago's greatest concentration of socio-linguistic and (purportedly) biological diversity (see Tzérikiantz 2008; Mondragón forthcoming). Its many distinctive physical traits include Vanuatu's highest mountain (Tabwemasana), longest river (the Jordan), largest protected harbour (Big Bay) and the

---

3   See Boomgarden (2007) for a similar approach in a neighbouring culture region.

Cumberland Peninsula (see Figures 3.1 and 3.2). Because it is the primary area in which most of the events recounted in this chapter take place, the following is a summary explication of the major sub-regions that surround and give form to Big Bay.

The west coast is the most distant and isolated part of North Santo, when compared to the southern part of the island and the town of Luganville, where Vanuatu's second-largest service sector and airport are located. The shoreline of West Santo is characterised by a very narrow set of beaches lying at the bottom of sharply vertical mountain ranges that tower over the ocean with an average height of more than 1,000 m and host some of the richest remaining areas of high-value timber in Vanuatu. This mountainous wall stretches north, all the way to the tip of the Cumberland Peninsula, and is a product of ongoing and intense seismic uplifts.[4]

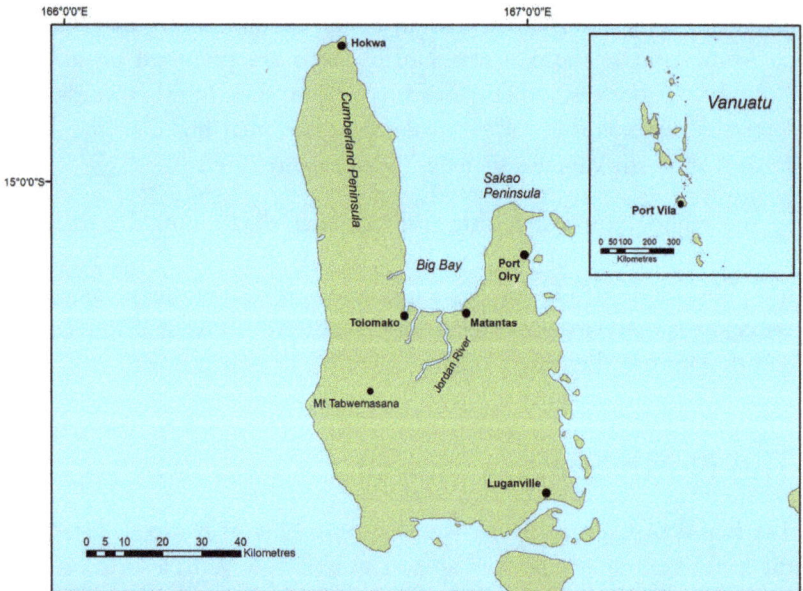

Figure 3.1 North Santo and the Jordan.
Source: Cartography by Jerry Jacka.

---

4   Santo lies across the junction between the Australian and Pacific tectonic plates.

Figure 3.2 The Santo waterscape, as viewed from the south.[5]
Source: Photo by the author, 2010.

---

5   Note that the Jordan is not even represented on this tourist billboard in Luganville, even though it is the largest body of water on the island. In comparison, the river that is represented here as running from north to south is in fact much smaller and shorter than the Jordan; however, it is becoming a popular destination for adventure tourism day trips, and that perhaps explains its prominence in the imagination of some of the townspeople.

At present, the west coast hosts a few small and scattered coastal settlements whose only quotidian access to Luganville and the south is through motorboat routes that are neither regular nor economically accessible to most local villagers. The most frequent runs are operated by boat owners ferrying agricultural produce to southwest coast villages whose inhabitants then drive these crops to the urban market. The extremely rugged interior of the west coast is dominated by the presence of Mt Tabwemasana, which, at 1,879 m, is Vanuatu's highest mountain, and the third-highest peak in the Pacific Islands outside of New Guinea, Hawai'i and New Zealand. The headwaters of the Jordan River originate in the high gullies surrounding Tabwemasana, an area that is isolated and very sparsely populated as a result of the forceful displacement of most interior villages to the west coast during late condominium times.[6] Jean Guiart wrote an account of this high region following his trek through several of its communities in the mid-1950s (Guiart 1958).

In contrast with the western half of the island, the shoreline of East Santo is characterised by relatively flat ground endowed with rich soils. For this reason it became one of the premium territories for the establishment of cattle farming and copra plantations during the colonial era. The transformations of this farming legacy survive to this day, but over the past decade this territory has begun to experience a new bout of change as a result of two factors: first, the indiscriminate, speculative and often illegal purchase of coastal tracts by Australian developers; and second, the completion, in 2009, of a two-lane blacktop built with funding from the Millennium Challenge Account, which was established and supported primarily by the United States Government. Consequently, the ni-Vanuatu settlements of the east coast are more numerous, more densely populated and readily connect to Luganville. This explains the predominance of several east coast communities within important sectors of the Luganville economy, most notably public transport and stevedoring.

Big Bay is a very large U-shaped formation that lies between the extended arms of the Cumberland Peninsula to the west and the much smaller Sakao Peninsula to the east (see Figure 3.3). The total maritime surface encompassed by Big Bay is considerable—one cannot see the ends of its peninsular arms from the southern end of the bay, and it takes eight hours

---

6   The British and French jointly governed Vanuatu from 1906 to 1980 under the authority of an agreement known as the British–French Condominium. Vanuatu became an independent country in 1980 (MacClancy 2002).

to traverse it lengthwise by sea, in good weather, along the most extended western shoreline below the Cumberland mountain range. The extended region of tropical forest that rests beyond the shoreline of the black sand beaches at the bottom of the bay is sparsely populated and watered by a meandering network of streams and rivers that variously feed into, and extend out from, the mouth of the Jordan River, which empties almost exactly at the centre of the cul-de-sac that forms Big Bay's interior end. The only access into this area is by two dirt tracks: one that cuts across 'bush' Santo from the south, and has been developed by various logging companies over the past 20 years; and the second that cuts across the eastern ridge of the Sakao Peninsula and connects with the East Santo road that leads down to Luganville.

Figure 3.3 The shore of Big Bay, facing the western, mountainous interior of Santo, as viewed from Matantas.
Source: Photo by the author.

Importantly, both tracks end at the southeastern corner of Big Bay, which is capped by the coastal village of Matantas. The only other road into North Santo is a dirt track that turns off from the Luganville–Matantas bush track, winds its way west across the Jordan and eventually forks again to reach the remote coastal settlements of Malao and Tolomako. Traffic along this western bush track is minimal since neither of those communities represents a major point for the transfer of goods or people. Beyond them, into the bush and the huge mass of the Cumberland Peninsula, there are no more roads. It is for this reason that Matantas constitutes the major point of entry and departure for vehicles coming and going from the south of Santo, and, conversely, for the motorboats that service the scattered communities of the western shore of the Cumberland Peninsula. Hence, Matantas features prominently as a crossroads for the movement of people and things (consumer goods, exchange items, tools, animals and many other objects) into and out of Big Bay. Not surprisingly, the people of this small community—less than eight extended families totalling around 200 permanent residents over the course of the past decade—have reaped ongoing benefits from their situation as middlepersons, as well as sea and land transport owners, in relation to other Big Bay and East Santo communities. As I explain in the final section of this chapter, the socio-economic prominence of contemporary Matantas, in relation to the rest of North Santo, is further enhanced by the awe that it also inspires through one of its powerful places: a freshwater spring and its spirit guardian, present in numerous local accounts relating to the origin and continuity of Santo Island.

Beyond Big Bay, the single most important environmental feature associated with the waterscape of North Santo, other than the Jordan River, is the extraordinary complex of terraces and canals that characterise large tracts of the steep mountainsides of the Cumberland Peninsula. These enormous irrigation works are indispensable to the proper maintenance of the staple cultivar of Cumberland Peninsula communities—namely, the water taro (*Colocasia esculenta*) (see Figure 3.4). As such, they are an impressive testament to the long and continuing history of major human modifications to the North Santo landscape and waterscape going back at least several hundred years. They also help to explain the prominence of wetland animals, water taro, waterfalls and natural springs (these last two being of critical importance to irrigation terraces) as entities closely tied to the origins of the river and to some of the principal North Santo lineages, to which I now turn.

Figure 3.4 Water taro near the village of Hokwa (Cumberland Peninsula).
Source: Photo by the author, 2008.

Stories of Origins and North Santo Topogeny

The most frequent exegetical reference that emerges in relation to the origin of the Jordan River is the *kastom stori* ('customary story' in Bislama)[7] that has become the most commonly accepted 'origin myth' for this waterway among the approximately 10 different language communities native to the North Santo region (Lynch and Crowley 2001: 45).

In brief, the Jordan is said to be made up of two smaller rivers—the Ora and the Lape (or Labe)—which exist as smaller, separate tributaries that come together at the upper end of its extensive watershed. The Lape is said to be the stronger of the two, and this appears to be proven by the large amount of trees that are swept from its banks and carried downstream by its powerful current. The following is a summary version of this *kastom stori*:

---

7   Bislama is a Vanuatu pidgin that functions as the lingua franca for all ni-Vanuatu.

In a previous time, when the wind blew through the trees and brought their sound to the sea below [in Big Bay], the salt water exclaimed in fright, 'Oh! this river wants to break through and dump its rubbish on me. I shall stop it.' It therefore held the Labe back by building up the coastline at the mouth of the Jordan with sand and stones. The Labe could no longer reach the sea.

Meantime, the other tributary, the Ora, was slow and hid the rumour of its waters by dampening it with the plentiful stands of *burao* [cottonwood] trees that line its banks. Upon seeing that the various mouths of the river delta were blocked up, it conferred with the Labe in order that they should join forces and break through to the sea. Hence the Ora, being the one who offered itself as the 'foundation' for their joint endeavour, remains the dominant one, and its waters are on 'top' all the way from where it joins the Labe and down to the bay, and this is why the Jordan is often simply known as the Ora. People make offerings to it, and recognise it as one of the primordial spirits that created and sustain Santo Island.

That is how both tributaries come together, as we see them today. Through their joint force they just manage to constantly carve out small outlets through the pebble-strewn shoreline of Big Bay, and empty out into the sea, which cannot hold them both back. However, the salt water continues to try, and this is why the Jordan is known to be an unstable, powerful presence [described as *pe mo la hou soko ri*, 'water that shifts all about'/'that does not keep its bounds', in the North Santo languages of Sakao and Tolomako], whose spirit can dig out *nambanga* [banyan] trees by the roots or go into the forest if it wants.

This origin story speaks to the notion that the river is the manifestation of multiple, supernatural presences, and is a truly extraordinary entity with a will and a way that are not always benign to humans. Today, there are several dozen small communities scattered across the lower and upper Jordan basin whose inhabitants relate closely to the river, and to the animals and physical features that characterise its overall environs. In order to begin to understand the significance and depth of the human–environmental bonds that link the inhabitants of these places with their watery surroundings, it is necessary to obtain a general outline of the principal kin groups by which North Santo people identify and position themselves in relation to their circumambient spaces. As we shall see, the close proximity of human genealogical patterns and experiences of the environment with the powerful, morally laden and living features of the North Santo waterscape becomes evident upon a cursory examination of the names and origin stories associated with the major kin groups and lineages of this region.[8]

The social organisation of all North Santo communities is premised on matrilateral cross-cousin marriage, and very closely mirrors the general pattern of two opposed halves, or moieties, by which almost all other north-central Vanuatu societies organise their kin dynamics.[9] These moieties can be referred to as *mama laen* (matriline in Bislama), but they are more commonly known by the local term *navugu*, which seems to be the same across at least four of the different language communities proper to Big Bay. In accordance with the structure of most other North Vanuatu kinship groupings, each *navugu* acts as a matrilineal kin group that, in turn, encompasses several 'family lines'—*famle laen* in Bislama, *natau* in Big Bay languages. It is the *natau* that constitute the specific lineages that act like extended but coherent parties within the context of the ritualised exchange of persons and territory that make up the essence of reciprocity in these islands. Importantly, these moieties and family lines are not simply floating categories for a prescriptive 'system' of kin naming and relating. Because they derive their existence from place-based ancestors and forces, they are mutually constitutive of land and the broader environment. In a sense, they *are* a part of the living milieu of the landscape and waterscape at issue, and it is important to understand them as one of the ways in which personhood and belonging are made manifest. This is especially relevant in light of the moral dimensions of behaviour that are carried across in the *kastom stori* that I relate and refer to in the rest of this section.

The names of the two major *navugu* that operate across all of North Santo, and even extend into West Santo, are Navugu Maliu and Navugu Ova. The first of these names makes reference to the *maliu*, a net of local design that is employed for catching freshwater lobster, and which represents one

---

8   The following overview of some of the principal Big Bay kin groups, origin stories and their various relations to animal and hydrological features of the North Santo environment was possible thanks to a number of explanations (some tremendously detailed) that were proffered to me by Billy Tavue of Matantas over the course of several visits to his home. These materials were also greatly enhanced during a unique conversation that I was fortunate to have had with four kind and knowledgeable ladies—Naomi Wargongon, Vera Alavanua, Eslie Guru and Roslin Iaruel—who hail from different settlements on the east coast of the Cumberland Peninsula facing Big Bay. I remain indebted to all of them, and other Big Bay inhabitants, for their expert assistance and generosity.

9   The Torres Islands, at the far northern extreme of Vanuatu, offer an interesting exception to this system of opposed matrilines (Mondragón 2015b).

of the significant objects by which North Santo people distinguish their *kastom* from that of other ni-Vanuatu communities. For its part, the name Navugu Ova refers to the *ova*, or white heron, which makes its home in the swampy terrain that characterises the mouth and the lower basin of the Jordan River, and also fishes in the shallows by the seashore, and is therefore often described as a 'bird of the sea'.

Each of these *navugu*, in turn, encompasses a number of different *natau* whose names are shown in Table 3.1.

Table 3.1 Some prominent kin groups of North Santo.

Navugu Maliu	Navugu Ova
Natau Marae (named after the saltwater eel). This *natau* emerged from, and is associated with, Natau Hoaka, which no longer exists, and was named after a flowering plant with red and white flowers.	Natau Via, named after the wild taro, and said to be an offshoot of Natau Wuhu Ova (below).
Natau Taupala, named after the action of securing a pig bone. This originated as an offshoot of Natau Haviha, which no longer exists in Big Bay but is still extant on the west coast of Santo.	Natau Peta, named after the water taro. Sometimes also referred to as Natau Mata, meaning '*natau* of the eye'.
Natau Hocho, named after the action of tearing the meat from the bone of the pig. Also originated as an offshoot of Natau Haviha.	Natau Sarau, the meaning of whose name has not been recovered.
Natau la Piho (or Natau Wupoe), which is simply named after the pig.	Natau Vukarae, named after the flying fox, and said to be an offshoot of Natau Wuhu Ova (below).
Natau Maliu, named after a net used for catching freshwater lobster.	Natau Wuhu Ova, named after the 'white heron'.

Source: Compiled by the author from various fieldwork interviews with speakers of Sakao, Tolomako, Vunapu and Piamatsina languages.

The names given to the *natau* in the above table are not arbitrary, nor are they simply 'metaphorical' or representative of 'totemic' associations. In fact, they almost invariably refer to key creatures or naming events that are present in the origin myths of each *natau*. Hence, almost all of the creatures or objects that emerge in these names (pig, eel, water taro, net) tend to be the points of origin of each respective *natau*, or are strongly associated with the entities from which they originate. More important, they are almost all associated with powerful watery places. The shared origin story of Natau Pala and Natau Hocho offers a good example of these genealogical and topographical entanglements between water and persons:

Long ago, there were two sisters from Natau Haviha who had two girls each. They lived by the side of the Piavohoho River [a major tributary of the Jordan, on the southwestern interior of Big Bay], but were only able to give their children cabbage and the leaves of trees to eat, for they had no animals and no hunting skills.

One day, crying, the taller of the two sisters reached Septulahia, a place where there is a pool with a hole in its stony bottom into which a big waterfall enters and goes underground. This is the stone house of a powerful *navarie* [a freshwater spirit]. To one side of the pool there was a tree blooming with leaves, but as the tall sister climbed it, the tree fell into the pool. You see, in a fit of maliciousness her shorter sister had cut the base of the tree. The tall sister fell into the stone home of the *navarie*, who was dwelling inside with his wife.

After some time, her two girls went out to look for their mother, whereupon the *navarie* appeared before them and told them that she was safe in his home. The spirit then went hunting for pig in order to give them proper food. However, being generous and powerful, the spirit shared the pig with both the tall and the short sister, as well as their offspring.

Since she was considerate, the taller of the sisters held the bones of the pig, with soft, delicious meat still clinging to it, for her shorter sister to pull off and eat, because she did not know how to eat bone marrow. When they finished eating, the *navarie*, having fed them and thereby become a father to them, decided that the taller sister would become the origin mother of a new *natau*, which it named Natau Pala, the *natau* of she who held the pig bone fast. The shorter sister, meantime, would become the origin ancestor for Natau Hocho, the *natau* of those who tear the meat from the bone of the pig.

This is the story of how those two *natau* emerged, and that is why the lands around Piavohoho and Septualahia are known to be the ancestral territory of these two lines. Today, many of the members of these two *natau* still live close to their places of origin, and can be found dwelling in the villages of Piamaleto and Pialuplup [in the southern part of the Cumberland Peninsula, facing Big Bay].

Lest the binary opposition of Navugu Maliu and Navugu Ova seem a clear-cut exemplar of a more fundamental 'structure' in Santo and North Vanuatu kinship, it bears remarking that the origin stories of the *navugu* of Big Bay consistently mention the existence of at least one other *navugu*, which is known to have existed and played a major role in the patterns of ritualised exchange between the *natau*, but disappeared in the recent past. The same is true for at least two or three *natau* that are known to have existed but are no longer extant. The lesson here is clear: kin relations are modelled on ideal, prescriptive types of reciprocity and exchange (of persons, land and objects), but the actual interlocutors of these processes—the *navugu* and *natau*—are contingent parts, rather than unyielding, ahistorical wholes. The people of Big Bay are well aware that their kin groups are subject to emergence and disappearance at different spatial and temporal junctures. In other words, while the binary exchange of people and land is modelled on dualistic oppositions, these oppositions are subject to the contingent, historically and morally inflected unfolding movement of persons and things in time and space. Thus, rather than thinking of them as unyielding structures, the binaries at issue are subject (quite often, in fact) to the subversions and modifications that accompany actual political, social and environmental negotiations. In that sense, these genealogical origin stories offer us a window into the values and behavioural prescriptions that contemporary islanders tend to evoke when expressing their relations to the things of the North Santo waterscape.

The ancestral human–environmental entanglements outlined in relation to important sites of the North Santo waterscape are further informed— or, following my chosen analytical frame, 'layered'—by the enormous historical and Biblical purport that the communities of Big Bay attribute to the name of the Jordan River, following its presence in Christian scripture and its consequent association with the imagined landscape of the Christian Holy Land. Moreover, in the experience and presentations of North Santo people, these Biblical associations are inseparable from highly charged political processes, wherein the contested Christian past of Santo has recently become a crucial element in contemporary islanders' strategies of political and cultural differentiation in relation to the provincial and national frameworks of state power within which they are also enmeshed.[10]

---

10   See Luque and Mondragón (2005), and especially Mondragón (forthcoming), for a detailed breakdown of North Santo representations of the contested Christian past.

A Christian Toponymical Vignette

In April 2012, during a long conversation with Cirilo Palo, an influential knowledge expert of North Santo, I was offered the 'customary' name of Santo Island. The *kastom* name of Santo, Cirilo declared, was Yeven. 'It sounds like "heaven", doesn't it?' he added, in a conspiratorial tone. 'What local language does this name belong to?', I asked.[11]

> We don't really know any more. It's an old name. But for Christians [i.e. most local people, including himself], it is a powerful name. When the first white men, the Spanish, heard this name, they naturally chose Santo as the place where they would introduce Christianity to Vanuatu. That's why they gave the name 'New Jerusalem' to the settlement that they established in Big Bay, and the name 'Jordan' to the island's big river. They realised this was a special land. A holy land. The great southern land of the Holy Spirit.[12]

During our conversation, *olfala* ('elder') Cirilo added that few people today remember the *kastom* name for Santo—and this makes it all the more powerful.[13] His toponymical revelation was unexpected; in 10 years of tramping around North Santo, I had never registered a *kastom* name for the island. That it was unexpected, however, does not mean that it was extraordinary. As I have since discovered, Yeven is not a secret or restricted name, to be kept from public conversation. Its principal quality seems to be that it offers an important, shared link to an imagined Christian geography. As Cirilo observed, it is a powerful name, and in this regard

---

11   Cirilo is a resident of Port Olry, a community of east coast Santo that came into being in the first half of the twentieth century as the result of a mass, planned migration from people in Tolomako, on the southwestern corner of Big Bay. This means that his mother tongue is most likely Tolomako, but he is also fluent in the Sakao language of Olry, and quite possibly one or two other local languages. With regard to the name Yeven, this was not the first time that I found myself trying to obtain a narrow, concrete meaning for what has proven to be a widespread but polysemic category across different communities of North Santo.

12   Cirilo was drawing from the common translation, rendered into Bislama from English, of the original name given to Santo by Quirós—'La Australia del Espíritu Santo'. See Luque and Mondragón (2005) for a detailed analysis of the acts of naming and appropriation performed by the Quirós expedition.

13   Not coincidentally, Cirilo is one of the few who can still evoke it, to full rhetorical effect — as I witnessed one day later, when we discussed it with a group of Santo Islanders who passed through his village on the way to a Presbyterian Church meeting in the south of Santo. When asked about the name, and whether they knew it or could elaborate on its meaning, they confessed their ignorance, but immediately came to the same association with 'heaven' that Cirilo had proffered, and were quite ready to entertain the thought that it partly accounted for the fact that Santo was the first place where Christianity was introduced into Vanuatu.

it constitutes a key term in the 'ever-shifting social geometry of power and signification' (Pannell 2007: 74) by which the people of North Santo 'make persons' and 'make place'.

In a similar vein, New Jerusalem is important because it was the name given by the first Europeans to intrude on Vanuatu, when they came ashore at the site on which the present-day village of Matantas was built. In the words of *olfala* Cirilo:

> The chief in charge of guarding this ground [Matantas and its surroundings], whom the Spanish killed long ago, he was a great chief. He had a name. No one remembers it now. But I do. Because it is my responsibility. He was known as chief Masorow. His matriline [*natau*] survives. His descendants are still here. (Cirilo Palo, personal interview, March 2010)

Given their close association with the first arrival of Christianity, a decisive event in most ni-Vanuatu reconstructions of the local and national past, it should not be surprising that names such as Yeven, Jordan and New Jerusalem represent a powerful cluster of toponyms through which North Santo people render their shared territory (as opposed to the ancestral territory of a particular matriline) as a unique space of spiritual and historical purport in relation to the southern part of the island and Vanuatu more generally.

Importantly, one of these toponyms does not refer to a 'grounded' part of the landscape (variously *venue, vanua, venie* or *fana*), but to a body of water (*nepe* in several local languages). By evoking the customary name for the River Jordan in the same breath with that of Santo, *olfala* Cirilo provided a useful reminder about why the circumambient world of North Santo cannot be adequately approached by understanding it principally as a 'landscape': the emphasis on 'land' eschews the subtle but powerful intertwined presence and effect of the various elements of the 'waterscape' that constitute the islanders' lived worlds.

Given that they are the primary sites of the waterscape in which the stories and people that I mention in this chapter live, it is convenient to turn again to the Jordan River, and more broadly to Big Bay, which is an extraordinary (in the sense of very large and unique) part and product of the watershed of the Jordan.

The Jordan Basin: Shared History and Watery Sites of Power

The River Jordan, like the *kastom* name for Santo, represents a key element of the layered topogeny, 'the recitation of an ordered sequence of place names' (Fox 2006: 8), through which the people of North Santo anchor and act upon their landscapes and seascapes. In terms of the 'moral dimensions' of the local waterscape that I have mentioned previously, the Jordan is recognised as an especially powerful marker for claims to spiritual (religious) and political authority—that is to say, for 'truth' statements regarding knowledge of the moral, spatial and temporal qualities of the networked social and spiritual elements that constitute the Christian aspects of the history and geography of North Santo.

The broader scenario in which this history and geography is rooted is Big Bay, which constitutes one of the three major component areas of North Santo that I described at the beginning of this chapter. As explained previously, these consist of the western and eastern coasts and Big Bay proper. Here, however, I would offer a different way of typifying this region, in order to better highlight the positioning of peoples in relation to the Jordan basin. Rather than speak of east and west, I would concentrate on the contrast between the interior or 'bush' Santo, in which the sources and upper tributaries of the Jordan are located, and the coast. This contrast is not based on a radical opposition, but rather on a complementary relationship. From a coastal perspective, Big Bay emerges as an extension of the interior, insofar as it gives form to the lower reaches of the Jordan watershed.

'Bush' Santo, as I define it here, encompasses the largely inaccessible and extremely rugged mountainous interior through which the River Jordan runs; its soils and territory are constitutive of the overall basin, which is characterised by an extensive alluvial fan that empties out into the flatlands and marshes that make up the various 'mouths' of the Jordan. This latter zone comprises a shifting delta of rivulets that empty out into Big Bay. Importantly, because the mouth of the Jordan lies virtually at the centre of the bay itself, it neatly bisects the shoreline into an eastern and a western sector, each of which represents an important sub-region by which local communities define their sense of geographical belonging.

The alluvial terrain carved out by the Jordan basin, through which 'bush' Santo gradually descends, is more broadly flanked, to the east and the west, by the two extended peninsular mountain ranges of the Cumberland Peninsula and the Sakao Peninsula, whose vertical contours jut out into the northern sea, giving Santo Island its peculiar U-shape. The inward-facing sides of this U-shaped mountainous terrain frame and enclose Big Bay proper, while the exterior, seaward-facing sides of the Cumberland range and Sakao Peninsula give rise to markedly different physical and social coastal environments, known commonly as 'West Coast Santo' and 'East Coast Santo'. When viewed in contrast with the waters and coasts of Big Bay, it is easier to understand that the histories and networks of interaction of these two coastal regions are different insofar as, in addition to their inland connections, they have also been strongly oriented towards the open sea and other coasts and islands. Theirs is indeed a different waterscape from the one at issue in this chapter.

The above typifications should be understood not as geographic mirror images of local social boundaries, but as aspects of the genealogical entanglements that simultaneously link and separate 'upper' from 'coastal' peoples, or 'inward' from 'seaward' communities. Part of what I am trying to convey with these contrasts is a sense of the rich and changing patterns by which the settlements and language groups of the lower Jordan extend into and overlap with those of bush Santo. One of the forms that these entanglements take, in contemporary North Santo experience, is a shared sense of regional solidarity, as informed by recent events relating to the imagined past.

In terms of their place in the broader picture of Vanuatu's national history, the shared political and economic point of view of 'bush' and coastal North Santo communities rests on their perception of their region as a victimised landscape. As I explain in another text (Mondragón forthcoming), this forlorn image of their shared territory is largely a result of many North Santo communities having been on the losing side of the Santo rebellion of 1980 (Beasant 1984). The subsequent marginalisation of North Santo from the flows of wealth, development and political power following Independence has motivated both a regional sense of victimhood and the development of highly assertive and self-reliant forms of cultural and territorial identity. Across almost every language group in Big Bay, people keenly oppose the solidarity of North Santo to the historical development and ongoing interests of the southern half of the island, and indeed of the Vanuatu state as a broader, encompassing entity.[14]

The proximate manifestation of the post-Independence state into which many North Santo Islanders feel they were pulled against their will is the provincial bureaucracy of Sanma Province, with its headquarters in the distant, urbanised and increasingly wealthy township of Luganville. At the same time, beneath this sense of shared regional identity lies a tapestry of fascinating, largely untold microhistories of movement, conflict, belonging and resistance that give rise to the patchwork of different human communities and local environs of North Santo. This patchwork is susceptible to a variety of approaches or descriptions, none of which can aspire to constitute a totalising map since it encompasses innumerable life stories and localities. Some of these could include the hillside or coastal communities of the Cumberland Peninsula, with their roots in water taro and the enormous irrigated terraces in which their local staple has thrived for hundreds of years. It could also include the families of bush Santo who moved from under the shadow of Tabwemasana and onto the coastal lands abandoned by Tolomako people in the first half of the twentieth century, during the period in which the members of this latter community migrated *en masse* to the new town of Port Olry. The North Santo tapestry also includes the scattered lowland matrilines that make their home near the swampy banks of the mouths of the Jordan: some of these settlements are recent, having been founded by converts to new Pentecostal and other Christian denominations in the 1970s and 1980s. Conversely, there are longstanding communities that came together over a century ago as a result of conversion. Such are the Anglican folk of Tseriviu, near the middle of the Cumberland Peninsula facing Big Bay, or the extended network of Catholic villages that dot the landscape north of Tseriviu, all the way to Péséna, and thence hook up with the local diocesan headquarters in Port Olry. Finally, this map of interregional diversity would also include the small, isolated village of Hokwa, at the very northern tip of the Cumberland Peninsula, whose language is quite unlike the other North Santo languages, but whose people exhibit mythical and linguistic affinities with the inhabitants of the Torres Islands, over 100 km away across the open sea.[15]

---

14   For more on the victimised landscapes and microhistories that I am evoking here, see Brooks et al. (2008) and Stoler (2013).
15   Hence my surprise when, during a brief visit to Hokwa, I discovered that I could understand their vernacular speech.

This shifting patchwork of interregional pluralities of language, kinship and territory is powerfully shaped by—and rooted in—the geography of the Jordan watershed and its neighbouring ecologies. From this physical perspective, North Santo emerges as a unique, rugged world that continues to be difficult to access. Beyond the village of Tolomako and the mouths of the Jordan there are no more roads. The mountainous terrain and the multifarious, watery landscape (pools, marshes, waterfalls, swift creeks) have both shielded and excluded many North Santo peoples from the 'development' of other regions. It has also fostered a strong sense of self-reliance that works as a marker of shared histories and belonging.

In the final paragraphs of this chapter, I focus on the micro-historical perspective of a prominent North Santo community, the village of Matantas. My choice of Matantas derives, on one hand, from its past and present importance as a socio-geographical crossroads, a site of relevance within the kin and exchange networks that criss-cross Big Bay. It is also a site of critical importance in terms of the imagined Christian past of Santo—and indeed of Vanuatu as a whole (Jolly 2006; Mondragón forthcoming). There is, however, another reason for the relevance of Matantas in the regional imagination—a reason that takes us back to North Santo accounts of power, landscape and water.

As registered in conversation with many different North Santo people, there is a common perception that Matantas is a place of great spiritual power and danger. It is a taboo place whose most important toponymical marker is a spring that gives rise to the small creek from which Matantas gets its name—*mata* ('eye') plus *natas* ('salt water'). This creek is said to be the overflow from a massive, supernatural source of salt water that holds up the island of Santo from the inside. It is opposite and complementary to the source of fresh water from which the River Jordan springs, further up in the rugged mountains of the interior. It is a small but permanent manifestation of the broader powers of the physical cosmos, which has survived earthquakes, migrations, tidal waves, the disturbances of colonial plantation activity and other recent ways in which Europeans have continued to impress their notions of nature and history upon the local landscape. As such, it is a site that demands to be respected, honoured and feared.

The spiritual guardian of the Matantas spring is an ancestral woman, an origin spirit of the kind who created the basic things of the world: water, stones and the first ancestors of every human lineage. She resides, in stone form, at the bottom of the spring; awake and alert to the goings on of the world. Over the past four decades, this guardian spirit has intervened several times in the affairs of Matantas people: during the founding of the modern village in the mid-1970s, when she haunted those whom she would not tolerate as residents of this taboo ground; and during the opening of the dirt track in 1977, when she murdered several of the local and foreign people who were involved in the construction of the road. That road was not meant to be, because it represented the beginning of systematic intrusions from the south, and Matantas people have continued to make offerings and special forms of supplication to the guardian spirit in order to placate her anger at this increasingly busy road. The spirit was also present more recently, when she haunted a Canadian agricultural engineer who came to Matantas to try to implement a Chinese-funded pilot project for introducing rice paddies into local horticultural practices.

Finally, it is likely that the mythical purport of its guardian spirit is entangled with contemporary ways of situating Matantas in relation to the broader genealogical history of North Santo. Here, again, it is worth quoting Cirilo Palo, given his widely recognised authority on the layered stories of the land and people of North Santo:

> Matantas is the common ground of every matriline in Big Bay ... It is the 'mama ground', the foundational territory of every matriline in Big Bay. We all originally came from there; the founding ancestresses of every matriline came from there. So [...] this place ... it is different from all others. You see a settlement there today, but in the past it was not inhabited. It was taboo ground.

Conclusion

In another text, I describe the 'living milieu' of North Vanuatu societies as 'a complex interweaving of land- and seascapes, personhood, and "topogeny" [that] is continuously imagined and reiterated through flows of persons and things as a self-productive process that [as Biersack notes] gives rise to a "life-world of praxis and not just signification"' (Mondragón 2009: 116). In this chapter, I have explored the River Jordan's lower catchment basin from perspectives that incorporate the interweaving of watery geographies and genealogies as multi-layered 'horizons of agency' (Biersack 1990), in which persons and environments are in constant motion and transformation. In this regard, my materials are intended as an extension of a broader set of reflections about the humanised landscapes and seascapes of Vanuatu.

I have focused on how fresh water in its various forms is entangled with genealogical networks of people, presences and storied places that make up North Santo. To this end, I presented an outline—the first of its kind in the local ethnographic record—of the relationship between moiety and lineage names and the animals and places of the waterscape of North Santo. I thereby explored how water gives shape to powerful spirit presences—presences that, as in the case of the Piavohoho River and the spirit of the waterfall at Septualahia, serve as reminders of morally correct behaviour and genealogical chains of belonging. But I have also highlighted the presence of even greater primordial spirits, such as that of Matantas, whose perceived power extends to the origin story of Santo Island as a whole and continues to exert a forceful presence in contemporary affairs.

Finally, I explored waterscapes in relation to culture contact and history. History here can take the form of ancestral topogenies, or the imagined Christian past in the present, as well as of contemporary senses of marginalisation in relation to foundational events in the nation's history. I have thereby produced one of the first multilayered outlines of the North Santo context as a region that is profoundly and powerfully defined in relation to its watershed. The consequent image is of a unique, insular watershed, whose multiple, interleaving strands remind us that rivers, and fresh water more generally, are also critical components of Melanesian and Oceanic ways of making people and places.

References

Baviskar, A., 2007. *Waterscapes: The Cultural Politics of a Natural Resource*. Delhi: Permanent Black.

Beasant, J., 1984. *The Santo Rebellion: An Imperial Reckoning*. Honolulu: University of Hawai'i Press.

Biersack, A., 1990. 'Histories in the Making: Paiela and Historical Anthropology.' *History and Anthropology* 5: 63–85. doi.org/10.1080/02757 206.1990.9960808

Bonnemaison, J., 1994. *The Tree and the Canoe*. Honolulu: University of Hawai'i Press.

——, 1996. *Gens de Pirogue et Gens de la Terre: Essai de Géographie Culturelle — Livre I*. Paris: ORSTOM Éditions.

——, 1997. 'Les Lieux de l'Identité: Vision du Passé et Identité Culturelle dans les Îles du Sud et du Centre de Vanuatu (Mélanésie).' In E. Bernus, J. Polet and G. Quéchon (eds), *Autrepart. Empreintes du Passé*. Paris: Éditions de l'Homme/ORSTOM.

Boomgarden, P. (ed.), 2007. *A World of Water: Rain, Rivers and Seas in Southeast Asian Histories*. Leiden: KITLV Press. doi.org/10.26530/OAPEN_376971

Brooks, J., C.R. DeCorse and J. Walton (eds), 2008. *Small Worlds: Method, Meaning, and Narrative in Microhistory*. Santa Fe (NM): School for Advanced Research Press.

Fox, J.J., 2006 [1997]. 'Place and Landscape in Comparative Austronesian Perspective.' In J.J. Fox (ed.), *The Poetic Power of Place: Comparative Perspectives on Austronesian Ideas of Locality*. Canberra: ANU E Press.

Guiart, J., 1958. *Santo (Nouvelles Hébrides)*. Paris: Plon.

Hess, S., 2009. *Person and Place: Ideas, Ideals and the Practice of Sociality on Vanua Lava, Vanuatu*. New York: Bergahn Books.

Jolly, M., 1999. 'Another Time, Another Place.' *Oceania* 69: 282–299. doi.org/10.1002/j.1834-4461.1999.tb00374.x

——, 2006. 'Unsettling Memories: Commemorating "Discoverers" in Australia and Vanuatu.' In F. Angléviel (ed.), *Pedro Fernández de Quirós et le Vanuatu: Découverte Mutuelle et Historiographie d'un Acte Fondateur, 1606*. Port Vila: Délégation de la Commission Européenne au Vanuatu.

Lane, R., 1971. 'The New Hebrides: Land Tenure without Land Policy.' In R. Crocombe (ed.), *Land Tenure in the Pacific*. Melbourne: Oxford University Press.

Luque, M. and C. Mondragón, 2005. 'Faith, Fidelity and Fantasy: Don Pedro Fernández de Quirós and the "Foundation, Government and Sustenance" of *La Nueba Hierusalem* in 1606.' *Journal of Pacific History* 40: 133–148. doi.org/10.1080/00223340500176368

Lynch, J. and T. Crowley, 2001. *Languages of Vanuatu: A New Survey and Bibliography*. Canberra: The Australian National University, Research School of Pacific and Asian Studies (Pacific Linguistics 517).

MacClancy, J., 2002. *To Kill a Bird with Two Stones: A Short History of Vanuatu*. Port Vila: Vanuatu Cultural Centre.

Mondragón, C., 2009. 'A Weft of Nexus: Geographical Identity and Changing Notions of Space in Vanuatu, Oceania.' In P.W. Kirby (ed.), *Boundless Worlds: An Anthropological Approach to Movement*. New York: Berghahn Books.

——, 2012. 'Entre Islas y Montañas: Movimiento y Geografía Cultural en Melanesia y el Tibet.' In C. Mondragón, P. Fournier, and W. Wiesheu (eds), *Peregrinaciones Ayer y Hoy*. México: El Colegio de México (Antropología y Arqueología de las Religiones 4).

——, 2015a. *Un Entramado de Islas: Persona, Medio Ambiente y Cambio Climático en el Pacífico Occidental*. Mexico: El Colegio de México.

——, 2015b. *Concealment, Revelation and Cosmological Dualism: Visibility, Materiality and the Spiritscape of the Torres Islands, Vanuatu*. Paris: Éditions l'Herne/College de France (Cahiers d'Anthropologie Sociale).

——, forthcoming. 'The Troubled Image of a Biodiversity "Hotspot": Conservation, Christianity and History in Santo, Vanuatu.' In J. Bell and J. Halvaksz (eds), *Naturalist Histories*.

Pannell, S., 2007. 'Of Gods and Monsters: Indigenous Sea Cosmologies, Promiscuous Geographies and the Depths of Local Sovereignty.' In P. Boomgaard (ed.), *A World of Water: Rains, Rivers and Seas in Southeast Asian Histories*. Leiden: Brill. doi.org/10.1163/9789004254015_004

Rodman, M., 1987. *Masters of Tradition: Consequences of Customary Land Tenure in Longana, Vanuatu*. Vancouver: University of British Columbia Press.

Siméoni, P., 2009. *Atlas du Vanouatou*. Port Vila: Éditions Géo-Consulte.

Stoler, A.L. (ed.), 2013. *Imperial Debris: On Ruins and Ruination*. Durham (NC): Duke University Press. doi.org/10.1215/9780822395850

Taylor, J., 2008. *The Other Side: Ways of Being and Place in Vanuatu*. Honolulu: University of Hawai'i Press. doi.org/10.21313/hawaii/9780824833022.001.0001

Tryon, D., 1996. 'Dialect Chaining and the Use of Geographical Space.' In J. Bonnemaison, C. Kaufmann, K. Huffman and D. Tryon (eds), *Arts of Vanuatu*. Bathurst (NSW): Crawford House Press.

Tzérikiantz, F., 2008. 'Sevrapek City ou La Courte Histoire du Groupe de Recherche "Forêts Montagnes-Rivières" (Expédition Santo 2006, Vanuatu).' *Journal de la Société des Océanistes* 126/7: 207–220. doi.org/10.4000/jso.3602

4. Unflowing Pasts, Lost Springs and Watery Mysteries in Eastern Polynesia

ALEXANDER MAWYER

Sweet and Bitter Waters

With attention to both local and extra-local processes of construal of the cultural place and significance of water on the landscape in French Polynesia's Gambier Islands, I query the sometimes uncertain character and cloudy nature of water in this part of the Pacific. The molecular constancy of the substance notwithstanding, it seems that water is not always what it was or even where it was in recent pasts. For instance, in Eastern Polynesia the seemingly straightforward and highly culturally salient contrast in the binary opposition between *tai* and *vai*, salt and fresh waters, may displace the need to address the current complexity of the situation in which fresh waters on Pacific islands now stand.[1] Famously binary cultural logics can mask significant cultural ambiguities and practical uncertainties (Feinberg 1980; Mawyer 2014). Similarly, an enduring and valuable focus on the immensely profound place of Ocean for these 'peoples of the sea' (Buck 1938b) and of 'salt' (Hau'ofa 1998),

---

1   In spoken Mangarevan, the everyday distinction is often between *vai* and *vai kava*, sweet as opposed to salty or bitter waters.

and for the social and cultural histories of these islands, their original settlements and subsequent inter-island spheres of transaction and exchange, may have obscured the anthropological need to examine fresh waters in Oceanic contexts.

Helmreich (2011: 18) sources an early commentary on anthropology's heightened and focal fascination with salt waters to Malinowski, who for his part characterises the ocean as wild, unruly, fundamentally dangerous and requiring magnificent cultural affordances, but land as domesticated and relatively tamed (Malinowski 1931: 634–5, 1992: 30). Works like Sir Peter Buck's *Vikings of the Sunrise* (1938a) echo such perceptions, as does his *Ethnology of Mangareva* (1938b), where fresh waters, including important springs, receive scant attention. Although both domains of primal waters deserve reconsideration in the context of the broad revaluation of the place and status of liquid nature in contemporary anthropology (Strang 2004; Helmreich 2011, 2014; Wagner 2013), the status of fresh waters in Oceania particularly calls for refreshed attention given the possibility that in some Pacific places, such as Mangareva in the Gambier Islands, the apparent transparency of this everyday substance may draw attention away from significant anthropological murk around contemporary experience.

With reference to the Gambier, this chapter inquires into the significance of known springs, intimately felt to be important, that no longer bubble forth, much less flow downslope to cultivated lands or the ocean. What might be made of lost waters whose spectral presence in chants or legends can produce a sense of the uncanny when compared to otherwise expert knowledge of *nuku* and *kaiga*, maternal lands and lands held as property? Similarly, how might dry watercourses, unswampy swamps, ruins of traditional irrigation systems and other sorts of complicated objects on the contemporary landscape further support a reinterpretation of water in the Gambier Islands, in a region imaginatively dominated by the ocean, and where river courses may be short, not particularly voluminous and may even be seasonal? Even moderate waters can hold a significant place on the landscape, and blockages or discontinuities in flowing water can reveal much about local understandings of past and present island natures and local culture.

Vai, fresh water, in static, non-mobile forms, complements the riverine character of the chapters in this collection that promote attention to flow and flux and the generally moving and transporting character of gushing waters along gravity's gradient to the sea. There are, of course, locally important rivers in Eastern Polynesia such as Papeno'o on Tahiti, Taipivai on Nuku Hiva, or Wailua on Kauai. Even in the small yet high Gambier Islands, the possibly misleading translation 'river' for the word *tairuavai*—which I would gloss as 'watercourse'—can be found in the published lexicon (Janeau 1908; Rensch 1991; Tregear 2009). However, I draw attention to *vai* not only in its coursing flow but also in its form of standing waters, waters gently welling from the islands' freshwater lens as *puna*, freshwater springs or as seepages at the vertex of some ridgeline's watershed. The foundational role of springs as sources of standing fresh water in Eastern Polynesia has been commented on in numerous contexts, including in myth, cultural metaphor, studies of socio-spatial organisation, and recently as part of the linguistic and anthropological turn to the concept of landscape (Burenhult and Levinson 2008; Cablitz 2008; Levinson 2008). However, passing time has eroded or otherwise obscured some of the cultural practices and understandings on which many existing comments on Polynesian springs rest. Moreover, the basic presence or place of springs and other waters on the landscape of these islands can be far from stable over historical time.

Over the course of my fieldwork, I heard numerous youths in a variety of contexts assert that *kupuna* ('elders' or 'grandparents') are guardedly jealous of their knowledge. This distrust of the older generation's willingness to share *atoga tika* ('true histories') was commonplace and, as I have noted elsewhere (Mawyer 2008), is among the most significant dimensions of Mangarevans' reflections on cultural transmission in the early 2000s. For instance, one afternoon, after school had finished, I was chatting with some young teens playing with a soccer ball (*keuui popo*). This was in front of the *'are ao* (town hall) at the very centre of Rikitea, Mangareva's primary village. I had been on Mangareva for some time and my general interests were well known. So, attempting to impress me with their knowledge of things traditional, these boys began to recount the story of the legendary Noe's funeral canoe, which had ploughed into the mountain above Rikitea, creating a channel for divine, life-giving water to flow down to the lands around which the village would be founded, and creating an opening into the sea (Buck 1938b: 22). As it turned out, not one of the five boys playing could remember details of this story. Indeed, the punchline of the story, that the primary watercourse for the island's main village resulted from the piercing of the land by Noe's canoe and the release of flowing water, was completely unknown to them.

That a legendary canoe had pierced the ridgeline above and the name of the hero (though not the fact that he was dead at the time) was what they knew. That there was a connection between that cleft in the ridge and the watercourse at the centre of the village, which was previously sometimes called by his name, was unknown to them. Moreover, sharing what I knew about this, and describing how I knew it, led to an interesting moment when several of the youths stopped a couple of their adult uncles passing along the *ara nui* ('primary road') on their way home and, after asking them to confirm the basic story, pleaded (in French): 'Why don't we know this story?' To which the two more senior men responded with a worried, uncertain shrug.

The fact that youths, when accidentally confronted with evidence of slippages in the facts of intergenerational transmission of cultural knowledge, experience more or less passing, visceral and certainly meaningful moments of anxiety was hardly surprising, though I was struck then and now by their projection of responsibility onto more senior generations. In an intriguing moment of parallelism, when I was working some months later with a group of very senior experts on place names embedded in old texts, I did find myself surprised. In a previous meeting, as part of a locally initiated group compiling a new dictionary, we had worked through a list of place names, including those of a number of springs, compiled by Ioani Mamatui, a well-remembered expert of the previous generation, and one of Peter Buck's primary informants at the time of the Bishop Museum expedition in 1934. Daniel Teakarotu, one of the island's largest landholders and agriculturalists and the grandson of Karara, another of Buck's primary informants and a recognised *poukapa* (cultural expert), had set out to find several of the springs that featured in important myths or historical practices associated with the island's *'akariki* ('chiefs').[2] The fact that a week later he was not able to say exactly where these springs were—that they eluded him or had literally disappeared from the landscape—was a source of powerful astonishment to the group, at least equal to that experienced by the soccer-playing youths. In more or less the same words, I heard the men of this working group ask why they did not know where the specifically chiefly waters known to be associated with nameable lands were, and how they could have lost this knowledge.

---

2 'Special things associated with the ruler were his house, assembly place, and well or spring of fresh water ... a drinking pool or spring (vai) was reserved for the king and had a special name. The king, Ohokehu, had a pool at Vaitina named Vaihi with a special keeper in charge' (Buck 1938b: 153).

That evening in 2003, and again six years later in the summer of 2009, we discussed several possibilities. Perhaps the successful plantations of hardy pines established by the territory's agricultural service since the 1960s, which have utterly transformed the face of the islands (Mawyer 2015), are responsible for disrupting the subterranean shape of the freshwater lens. Or perhaps it was the massive erosion, both visible and regularly discussed in everyday life, caused by the wanderings and ruminations of the goats introduced in the nineteenth century, and the cows introduced in the twentieth century, on the upland slopes of all the archipelago's high islands. Or perhaps it was an increased human drain on the aquifer by mechanisms not entirely easy to calculate. Most worrisome, as suggested by the most soulful of the three senior speakers I was working with that night, perhaps the water could no longer be found because they had forgotten to honour the places in which it flowed, pointing also to the sacredness of the chiefly residential context of the sites of the former springs.

On reflection, it is not surprising that youths in the Gambier Islands or across Polynesia are anxious about cultural loss, but it is perhaps very surprising that *kupuna* and *koumotua* (honoured senior persons), with respected expertise and familiarity with linguistic and cultural shifts over their lifetimes, can be blindsided so profoundly by the common fact of water flowing or not flowing across the landscape. Such anxieties reveal that sweet waters (*vai*) can sometimes be bitter (*vai kava*), unusable like the sea, disrupting local sureties and senses of place.

In the sections that follow, I engage with the culturing of water on the landscape and identify some of the ways in which *vai*, as flowing or unflowing water, proves to be a complicated object of anthropological analysis on one hand, and of contemporary Mangarevan cultural experience on the other. Next I consider an anthropological anxiety present in scholarly texts regarding the sustainability of freshwater supplies in these islands. I then survey the rich and extended freshwater vocabulary of Mangarevan and related Polynesian languages, before returning to the question of Mangarevan seniors' reflections on lost springs, blocked flows and the ways in which waters provide or do not provide a living relation to the past. I conclude with a discussion of how water in Mangarevan experience relates to water as an object of anthropological inquiry.

Backgrounds and Foreshores

In French Polynesia's Gambier Islands, contemporary Mangarevans have a disconcerting relationship with their island environments, wild and domesticated alike, as I have begun to describe elsewhere (Mawyer 2015). The islands rise from the oceanic deeps, 1,700 km southeast of the political and social centre of French Polynesia on the island of Tahiti. Eight primary high islands, amounting to 27 km² of land surface, and a vibrant coral lagoon are inscribed in a barrier reef of around 90 km in circumference. The non-trivially varied micro-ecologies today include high grassy peaks, upland pine forests, traditional and neo-traditional downslope forests and highly varied near-shore house gardens and plantations with introduced, endemic and some other remaining indigenous flora in regular cultivation. Temporary watercourses can be found in most of the valleys. Remains of past irrigation stoneworks and terracing can also be found in many of the islands' bays. Traditional village sites tend to be associated with 'swampier' but cultivable areas in the foreshores of coastal bays, though today most of these sites are abandoned, having experienced constantly decreasing occupation since the mid-nineteenth century. French nuclear testing on nearby Mururoa and Fangataufa in the 1960s and 1970s led to a concentration of the remaining Mangarevan population in the main village of Rikitea, with a small number of families maintaining home sites in Taku and a few other locations (see Figure 4.1). Sand islands (*motu*) on the fringing reef are home to well-established though somewhat under-maintained coconut plantations, but today these islands are more likely to serve as the sites for weekend family vacations or 'get-aways' than for sustained occupation. Prior to contact with Europeans, all the islands of the chain were densely occupied, as were the sandy barrier islands that encircle the lagoon and the outlying atoll of Temoe to the east of the group. Only the three islands of Aukena, Akamaru and Mangareva were inhabited between 1999 and 2009, the years between my first and last visit. Other islands in the group were home to a few families until the 1960s, while Aukena and Akamaru themselves are home to only a few families, with temporary stays by labourers working on adjacent pearl farms.

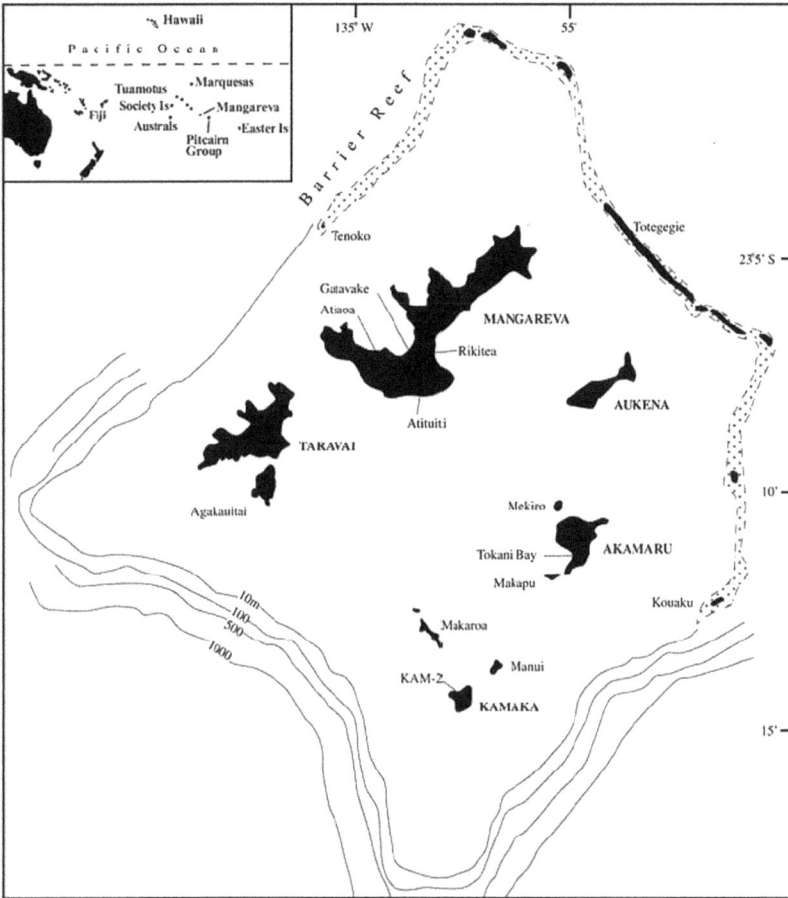

Figure 4.1 Map of the Gambier Islands in French Polynesia.
Source: Adapted from Anderson et al. (2003: 120).

In recent years, the Gambier's growing population has been reclaiming formerly inhabited spaces, yielding encounters that can be both surprising and disconcerting in their historical inflections as families clear land for the first time in many decades, or even in a century.[3] Densely tangled woods might contain a traditional house platform, boundary-marking

---

3   After several tragic early epidemics, careful Catholic mission records of the population began in 1838, and French state records began in the 1880s. In these sources, the population was recorded as being 2,141 in 1838; 2,400 in 1844; 2,270 in 1845; 1,350 in 1860; 936 in 1871; 656 in 1880; 446 in 1885; 508 in 1892; 580 in 1897; 529 in 1911; 501 in 1926; and 490 in 1936 (Vallaux 1994: 89). The population passed 600 persons for the first time in over a century in the 1990s. It was 566 in 1971; 556 in 1977; 582 in 1983; 620 in 1988; and 1,087 in 1996. In 2012, the population stood at 1,239 persons, with the expectation of continued growth (ISPF 2015a, 2015b).

sitting stone or stonework structure of unknown utility, or one or more charming quarried stone houses, in the French provincial style, from the height of the Catholic instauration in the mid-nineteenth century, long abandoned and often forgotten, alongside the stone ruins of earlier cultural complexes. In a sense, the contemporary landscape is saturated with the past. However, encounters with the material facts of historical displacements and discontinuities are not always dependent on there being anything there. Indeed, encounters with what is not there, as in the absence of things one might otherwise expect to see, can be just as disconcerting and can lead contemporary Mangarevan landowners to ponder uncanny questions of cultural loss.

For instance, cultural site inventories taken in 2001 by the Association Richessses du Fenua (a Tahiti-based cultural association), and redone in 2002–03 by the Association Reo Culturelle Mangareva, identified a startlingly small number of significant freshwater features of the landscape. Moreover, two of the six identified features were the result of the Catholic mission's work: wells dug near mission structures on Aukena and at Rouru, the site of the first Polynesian convent on Mangareva. Meanwhile, only two sites were identified from among the named springs or places associated with fresh water. One was the Bagnoire de la Princesse, a notable bath-shaped rock feature that is said to be imbued with sacredness and reserved to the now disestablished chieftainship, and is located on the Chemin des Vierges (also called the Chemin des Soeurs), the primary walking route to the now disestablished, ruined and vaguely haunted convent site at Rouru. The other was the Poatu Okame, also close to Rouru, a 'natural' reservoir that is said to have traditionally never dried out, even in the most extreme droughts, and that was thought to have curative powers in the pre-contact era. Another site is the break in the ridgeline called Manu Ka'u, where Noe's mythical canoe pierced the island, releasing a flow of water that descended towards the *marae* named Te Teito, a traditional Mangarevan temple, and whose course, sometimes filled with water, can readily be found at the centre of Rikitea village. Neither survey identified any of the previously notable freshwater springs whose names were sometimes given to the adjacent villages, or that served as the sites—if not exactly the sources—of chiefly authority (Buck 1938b: 6; Mamatui n.d.a).

Curiously, disappearing springs and absent flowing waters are in conversation with what Buck (1938b) describes as a myth of the oldest human times on these islands. Mangareva, the largest of the islands, is said to have been first discovered by two brothers, Moa and Miru, whose exploratory voyaging brought them to a delightful, verdant and fecund land. On arrival they discovered a magical spring guarded by two spirit beings who took the desirable form of beautiful women from another world called Po and for whom the spring served as a gateway. One of the spirit women was captured and agreed to live with Moa as his wife, while the other escaped back to Po through the spring-cum-gateway. After various and somewhat violent negotiations, Moa's new wife agreed to travel back to Po, secure numerous delicious and productive foodstuffs and return them to Mangareva in order to feed her husband and his brother in perpetuity. On her return with numerous fresh plantings and cultivars, she cautioned the brothers not to overindulge their gross appetites lest some unspecified ill befall them. Being human, the brothers understandably failed to heed this warning. The result of their overconsumption was the degradation of the sweet and savoury foodstuffs, most of which became bitter, indigestible or poisonous. Moreover, the magical gateway to the other world was blocked. Mangareva and the other islands were thus cursed with a persistent scarcity of sustaining springs and foodstuffs, and Miru and Moa returned to their own land (Buck 1938b: 20–21). Though Mangareva and the other islands were eventually settled from distant 'Iva, the remarkable implication of this Mangarevan myth is that the human presence on the islands was already responsible for the loss of a golden age of watery benevolence.

Later stories recounted in Buck's *Ethnology of Mangareva*, and in the so-called 'native manuscripts' that he relied upon (Mamatui n.d.b; Tiripone n.d.), reinforce this sense of a persistent crisis of the waters and the cultivation systems that rely upon them. Similarly, Anderson and colleagues observe that:

> The internal dynamics of Mangarevan cultural history are also likely to be of considerable interest in their own right. Hiroa's classic 'salvage ethnography' (1938) and the 19th century missionary account of Honoré Laval (1938), from which Hiroa derived much information, describe a society which displayed signs of a classic chiefly hierarchy severely challenged by resource limitations. (Anderson et al. 2003: 119)

One commentator among the members of the 1934 Bishop Museum Expedition described Mangareva as 'desolated' (St John, cited in Gregory 1934: 57), a comment that resonates with those of Buck. Buck has two passages that speak to the limited presence of water on the landscape:

There are no permanent streams, though some stream beds contain a trickle of water, and a few springs exist. The name of the freshwater eel (tuna) occurs only in myth. Seepages on the uplands or near the coast were utilized to the fullest extent for growing taro. From the sea, the islands have a rugged, dry appearance, for the cultivable land and forest trees are confined to the lower level of land stretching back from the bays. (Buck 1938b: 4)

And:

In Mangareva there are no large permanent streams, and the large irrigation scheme of central and northern Polynesia was possible on only a small scale. Irrigated terraces were formed in narrow strips below rock outcrops or cliffs below which water seeped out. Many in the Atituiti district consisted of three or four terraces but a few feet wide. In olden times every trickle of water was utilized, and disused terraces are to be seen high up on the hillsides. Laval states that the soil in such mountain terraces had to be frequently replaced. In a cultivation examined, a spring of water issued from below a low cliff on the hillside and in the course of time had cut down a rocky water course. A channel had been cut at the source of the stream to lead the water some yards downhill to the first terrace cut out of the side of the hill with the outer edge built up with stones to form a retaining wall (*kala*). The main channel was termed a *tairua*, which is also the general term for a stream or channel. Two smaller channels (*ka'iraga-vai*) were cut toward either end through the raised outer edge of the first terrace to lead the water down to the second terrace formed like the first. The small channels were blocked with earth or grass to flood a terrace when required, but the channels were sufficiently high above the main level to keep the terraces wet. The channels carried off the overflow. A third terrace completed the series, and a side channel carried the overflow back into the old stream bed. (ibid.: 226)

More recently, since the turn of the new millennium, Mangareva and the other Gambier Islands have been the site of extensive archaeological work that appears to confirm this understanding of a mixture of freshwater resources in the landscape (Green and Weisler 2000; Kirch et al. 2010; Conte and Kirch 2011; Anderson et al. 2003). For instance:

Rikitea offers a sheltered bay and canoe landing, along with one of the largest valleys with many freshwater springs at the base of Mt Duff that are watering a swampy alluvial basin which in historic times contained the most important zone of intensive taro (*Colocasia esculenta*) cultivation. (Anderson et al. 2003: 123)

But at the same time, the island chain as a whole suffered from a deficit of dependable freshwater springs, running waters and swampy grounds (Green and Weisler 2000, 2002), notwithstanding significant average annual rainfall of approximately 2,000 mm (Vallaux 1994: 35).

There is a quite remarkable variability in the simple material fact of fresh water in this island group. Precisely established survey results conclude that there are no dependable water sources here, while simultaneously drawing attention to wet, swampy, spring-filled and irrigated lands that have supported a population of some size for more than a millennium (Kirch 2007; Kirch and Rallu 2007).

Inheriting a tradition of interpretation from Te Rangi Hiroa (Buck 1938a) and Sahlins (1958), Kirch (2007) and Conte and Kirch (2011) reinforce a sense that these islands constitute an unusual environment when compared to other high islands elsewhere in Eastern Polynesia (Conte and Kirch 2011: 253).[4] Their eco-social model focuses on the centrality of human impacts on the transformation of the terrestrial environment, as exemplified by the diminishing abundance of terrestrial snails and certain bird species as markers of forest health, and points to 'increased deforestation and extinction of gastropods in the late period' (Conte and Kirch 2011: 260).[5]

An early and long-running erosion sequence (Anderson et al. 2003) is associated with the argument in favour of a four-stage transformation of the local ecology:

- an era of native flora and fauna that preceded the first human (Polynesian) settlement
- the era of early settlement, archaeologically argued to have been reasonably dramatic in terms of environmental impacts, including the depredations of introduced mammals and the broad employment of cultivation practices in the domestication of the archipelago, leading to a massive retreat of 'forest' to high slopes and areas unsuitable or undesirable from the point of view of cultivation

---

4   The use of the Gambier case as the exception that proves the model's rule also has an intriguing history in other contexts, including political order and social stratification (e.g. Buck 1938a; Sahlins 1958).

5   Conte and Kirch also speculate that the decrease in indigenous bird populations, whether due to human resource use or the activities of *Rattus exulans* (the Polynesian rat), perhaps accelerated deforestation through nutrient depletion due to diminished enrichment from excrement.

- the historical era, with the introduction of additional grazing animals, specifically goats and cows, which led to the final demise of 'native forests'
- the massive replanting of many upland and lowland terrains with new, generally exotic woods, and the general eradication of cows and goats from major islands (Butaud 2009; Thibault and Cibois 2011).

Interestingly, from the point of view of fresh water, *vai* takes on a curious figuration in these accounts of island ecological transformations, eco-social interactions and the complex relationships between multiple species and features of the archipelago's ecology. Some attention is given to weather patterns, and occasional multi-year droughts predicted by the model, but little attention is given to the transformation of the presence of fresh water as root systems are disturbed and erosion alters the face of the landscape, or as grazing mammals trample upon shallow basins or reconfigure watercourses.

The Semantics of Water

The landscape is not the only entity under erosion in the contemporary Gambier world. Mangarevan, an Eastern Polynesian language, is also severely endangered (Charpentier and François 2015: 119–20). Attention to the lexical semantics of water may further extend a sense of the character of fresh water in the here and now. Aside from several highly technical discussions in historical linguistics (e.g. Dyen 2005), the idea of 'water' has not been thoroughly scrutinised in Austronesian linguistics and the culture-historical analyses that become possible through lexical analysis and the comparative method (Blust 2013). However, in a remarkable and recent work on the 'hydrological' lexicon in Jahai, a language of the Malay Peninsula, Burenhult (2008) provides a strong analytical framework for analysing the 'structural and semantic properties' of water at the intersection of language, culture and cognition. In Eastern Polynesian languages, including Mangarevan, several cultural distinctions are readily visible in comparative contrast with other language families.

As clearly evident in Greenhill and Clark's (2011) POLLEX data,[6] water provides a wonderful example of the proximity of these closely related languages in Eastern Polynesia, as well as evidence of the

---

6   POLLEX is a comparative lexical database for Polynesian languages.

remarkable success of ancestral navigators in transporting language and culture across vast oceanic spaces (Table 4.1). For speakers of the proto-language(s) of the navigating ancestors who settled these islands, and for contemporary speakers of Eastern Polynesian languages, 'water' was and is not a conceptually simple or singular 'fundamental category' (Burenhult 2008). Rather, across these cultures and languages, water is a fundamentally bifurcated category. Whereas speakers of various European languages might conceive of *sea* water and *fresh* water as two aspects of the same substance, Eastern Polynesian speakers might conceive of *sea water* and *fresh water* as two different substances.

Table 4.1 Primary lexical forms of water in Eastern Polynesia.[1]

Form	Meaning	Hawaiian	Marquesan	Mangarevan	Tahitian	Tuamotuan
Tahi	shallow sea near shore or in lagoon, salt water	Kai	Tai	Tai	Tai	Tai
Wai	fresh water	Wai	Vai	Vai	Vai	Vai
Waiho	leave, lay down, place	Waiho	Avai	Vei'o	–	Vaiho
Wai-keli	well, water-hole	Wai-'eli	–	–	Vai-'eri	–
Wai-rua	spirit	Wailua	–	–	Wairua	Vaerua
Wai-marie	calm, placid	–	–	–	–	–
Wai-mata	tears	Waimaka	Vaimata	–	–	Vai mata
Wai-puna	spring (of water)	Wai puna	–	Vaipuna	Vaipuna	–
Wai-tafe	stream, river	Waikahe	Vaitahe	–	Vai tahe	–
Wai-uu	milk	Wai-uu	Vai-uu	–	Vai-uu	–
Wai-wai	weak, watery	–	–	–	–	–
Puna	bubble or well up (of water); a spring	Puna	Puna	Puna	–	Puna
Pugaehu	bubble or well up (of water); a spring	–	–	–	–	–

[1] Blank cells indicate the absence of verified terms in the lexicons of the different languages, even where the basic lexical form makes sense to local speakers.

Source: Greenhill and Clark (2011).

Beyond these initial paradigms of water comparable across closely related Eastern Polynesian languages, a catalogue of all terms in historical Mangarevan whose semantic field encompasses fresh water reveals further aspects of the culturing of water (Table 4.2). These include: (i) landscape features (*puna*, *tairuavai*) as fundamental categories; (ii) descriptive terms for qualities or characteristics of water(s); (iii) the moving or flowing action(s) of water; (iv) action relative to landscape features; (v) actions of fabrication relative to managing waters; (vi) actions of manipulation; and (vii) metaphorical or broadly figurative terms (e.g. lies are 'words like water').

Table 4.2 Lexical inventory of fresh water in Mangareva.

Form	English Gloss
'Akatairua	Create a trench or watercourse.
'Anautama Vaikino	House for women to repose or retire to during menstruation.
Ka'iraga Vai	Small irrigation trench within a plantation or garden.
Karekare Vai	Movement on the surface of a water.
Koko	To break, said of billows and waves; an egress; issue; way out; to run without overflowing or fall.
Mi'ami'a	Partially frizzled, inclined to be crisp, said of the hair, but also of a wave or billow.
Patito	To leap over anything; to jump a brook.
Puna	A source, a spring; to boil; pupuna, plural of the subject.
Punaga	Boiling.
Punapuna	To emerge, as a spring or fountain does; to spit or jet out, as water.
Punavai	A spring of water; a fire whose flames ascend.
Rupou	To drink by stooping down at a brook; to put one's lips to the water in a cup. Rupoupou, plural of the action; rurupou, plural of the subject.
Ta'ega	Marks made by flowing water; the bed of a brook, etc.
Taetavai	Water spilt or shed naturally, but not forming a brook.
Tairua, Tairuavai	River, stream, irrigation trench, cf. ka'iraga vai.
Tatapiga Vai	Action of bailing water out of a vessel.
Toki-Vaitau	Abundance; affluence.
Unuga-Vai	A drink, a beverage; the action of drinking (unu is to drink).
Vai	Water.
Vaiake	The Deluge.
Vaie	Water that lies stagnant in places otherwise dried up.
Vaie'u	Water that has been stirred up.
Vaiea	Water that oozes out from underground.
Vaierei	Water in a coconut (ere'i or tumuere'i are the coconut)

Form	English Gloss
Vaieroero	A prattler, a babbler; a preacher; to preach; to propagate a doctrine, true or false.
Vaieroeroraga	Propagation of lies.
Vaikava	Bitter water; alcoholic beverages; to be ungrateful, forgetful of past kindness.
Vaikavaraga	Ingratitude.
Vaikura	Red water.
Vaimaga	Water that has its source in the mountain.
Vaimegeo	Vinegar.
Vaioko	Ice.
Vaipaoko	Fish pond, specifically used for poisoning fish.
Vaipu	A pond.
Vaipuna	Water that springs from among stones; a muddy source.
Vaipuruka'a	Water from coconut butter.
Vairega	Water tinged with a yellow colour.
Vaitai	A mixture of salt and fresh water; brackish water (tai is sea, salt water).
Vaitapu	Holy water.
Vaito	Sugar; water of the sugarcane.
Vaituragi	Water from a beneficial rainfall.
Vaivai	Moist, humid; Semen humanum.

Sources: Janeau (1908); Rensch (1991); Tregear (2009).

In Mangarevan, as in closely related languages across Eastern Polynesia and Oceania, fresh water historically evidences a semantic lushness in its highly variable deployment across human contexts, whether literally or metaphorically. However, flowing water does not appear to be an ontologically robust domain in Mangarevan,[7] at least when compared to English (river, stream, brook, creek, run, ditch, rivulet, and so forth) or Jahai, an Austro-Asiatic language traditionally spoken by hunter-gatherers for whom the riverine character of their landscape is among the most salient of ecological features (Burenhult 2008). Half of the terms shown in Table 4.2 may now be as obsolete or unknown to speakers under the age of 50 as the physical locations of many of the waters to which they might once have applied.

---

7   Cablitz (2008) identifies a smaller list of primary water-feature landscape terms in the Marquesas, with a comparable suggestion that freshwater features did not constitute a robust semantic sub-domain.

Culture History Distilled, Purified?

Eastern Polynesian peoples are not and have never been only *tagata tai* ('peoples of the sea'). They have also always been *tagata vai*, entangled in the earliest histories of their places, ancestors and communities, reaching back to what Serge Dunis (2006) called the 'sweet and watery root times' of Polynesian cosmological and eco-social beginnings. These islands, like the much larger islands of Melanesia, with their famously numerous and sinuous rivers, deserve refreshed attention to their waters in several distinct and distinctly overlapping domains, including the material, economic, legal, political, ontological, mythological, toponymic, linguistic and eco-social. In this and the concluding section of this chapter, I prefer to emphasise the overlapping quality of these domains, drawing attention to the problems inherent in the construction of water as an object of scientific investigation.

In reflecting on the various contexts of two ethnographic anecdotes presented at the outset of this paper, expressing the concerns of youths and their elders alike about historical blockages and the possibilities of their removal, I suggest that the blocked transmissions that characterise my understanding of these anecdotes seem to resonate across different dimensions of *vai*—both in these islands and perhaps elsewhere in the parts of Eastern Polynesia where I have lived and that I know well.

While it is clear that some historically recorded water terms are no longer intelligible to contemporary Mangarevan speakers, and others do not mean quite the same thing even when they are, it is also clear that Mangarevan youths are not receiving all of the past culture of water even when there is little doubt of the semantic content of a term. In the cosmological, mythical or legendary domain of heightened speech genres and texts, the recession of the stories of the past into historical shadows and uncertainties mirrors the complex legacies of place names now felt to have been previously meaningful, concrete and present. The disruptions include the knowledge of irrigation systems and agricultural transformations of the terrain, as well as the identification of particular named bodies of water with specific kin groups or chiefly residences. Meanwhile, perhaps as both a symptom and cause of this epistemological and ontological murk, the legal regimes regulating water and its flow, or its nature as the subject of property rights, have a particular history dating from the era of the Mangarevan kingdoms of the Ariki Nui through the modernisation of the Code Mangarévien, the legal regimes of the French

Protectorate, and then into the present. Finally, of course, the actual waters on a small island are not entirely dependable in quantity or quality, location, efficacy (for irrigation) or potability (for consumption).

At the same time, locally rooted persons (*tagata tumu*) are hardly the only persons to now display uncomfortable, ambiguous relationships to island waters and their significance in circulating, sometimes boiling, conversations about islands as ecological laboratories. In the Gambier Islands and on Mangareva, the historical and prehistoric status of water has been put to work in a potent analysis of eco-social collapse (Diamond 2005).

Having lived on these islands, I find it easy to be profoundly fascinated by the recent efflorescence of archaeological research on their environmental and cultural history, including its attentiveness to deforestation and ecological degradation. At the same time, I worry about some disquieting dimensions of the transformation of hard-won archaeological, biological and environmental knowledge into moral fable. Indeed, the circulation of a narrative (Diamond 2005) that casts a moral colour over Mangarevans' actions and agentive responsibility for ecological and social 'collapse' over the centuries since their arrival suggests a problematic directionality to ecological change. Diamond's argument that Mangarevans lacked a management culture sufficient to care sustainably for their islands displaces more odious dimensions of their broader history—notably the fact that Mangarevans were in robust possession of a highly stratified and complex Polynesian cultural practice at the time of European arrival, as well as the role of aggressive and sometimes sinister European cultural interventions in the process of ecological change, if not the literal collapse, of some aspects of the island biogeography at that juncture (see McAnany and Yoffee 2010).

I remain convinced of the need to add voice to concerns about the appropriateness or accuracy of 'collapsatory' accounts of Pacific peoples and their cultures in a sea of vibrantly Polynesian islands struggling with the post-colonial legacies of odious European imperialism, including (in the Gambier case) the long-term consequences of nuclear fallout (CESCEN 2006). I raise the spectre of these concerns because I am conscious of the possibility that attention to blocked waters and unfulfilled flows, to changes in the character of the landscape at the intersection of frothy nature and instrumental culture, risks reproducing what would be for me an unsatisfactory narrative and unwarranted historical causality—

the reduction of historical experience or meaning to a model in which increased human impacts on the landscape deterministically yield a diminution of natural resources and, through transformative eco-social feedback loops, an increased disruption in culturally foundational human practices.

> Like all historians, we configure the events of the past into causal sequences—stories—that order and simplify those events to give them new meanings. We do so because narrative is the chief literary form that tries to find meaning in an overwhelmingly crowded and disordered chronological reality. When we choose a plot to order our environmental histories, we give them a unity that neither nature nor the past possesses so clearly. (Cronon 1992: 1349)

In this celebrated account of eco-social transformation, Cronon is making a point that I take to be quite applicable to Mangareva.

With respect to the island's waters, the past does not flow into the present through some process of osmosis, as in a gradual accumulation of facts or knowledge across the membrane of time, a linear historical process that affords a reductive historical moral (e.g. Diamond 2005). Rather, the culture history moving through Mangareva's waters is also operating by means of a reverse osmosis whereby the porous membrane of time sometimes allows for a backward flow of the concrete, the known and the knowable. Under the action of the solvent of the present, the stuff of culture history—such as the names of sacred springs, their relation to cultural transmission or collective morality, or the relation of the founts of chiefly power to the liquid ties that bind a community to landscape—sometimes moves from the visible and flowing back into a world that is subterranean and difficult to fathom. This happens when water sources are lost, even though the names remain, or when the language-encoded meaningfulness of water is lost, even though the waters themselves are still present. Meanwhile, recently significant levels of annual rainfall obscure the traces of putatively dry pasts, even as it is now difficult to locate the springs and irrigation channels that made life possible within those times.

Conclusion: Unflowing Present-Pasts or Past-Presents?

At the same time as pointing to the Mangarevan experience of water as indexical of a broader array of historical signs and counter-signs, a broader concern with the slipperiness of seemingly uncontrovertible 'natural kinds' in established anthropological discourses seems to be warranted. Contextualised by anthropology's long entanglement with the human condition at the intersection of nature and culture, the recent literature in ecological, eco-social, multi-species, ontological and other collective shifts or currents in scholarly production indicate the existence of what Kuhn (1970) called anomalies in normal science. Are rivers and other fresh waters, in Pacific island contexts, anomalous scientific objects that disturb, or even violate, 'expectations that govern normal [anthropological] science' (Kuhn 1970: 53)? As Daston notes in her introduction to a very fine collection on the emergence, submersion and evaporation of objects of disciplinary inquiry:

> Scientific objects may, like dark continents and invisible planets, take centuries of theoretical and empirical effort to find, or be accessible only by means of the most powerful instruments, but in their essence they are as enduring as quotidian objects (Daston 2000: 2).

Recently re-reading her introduction, I was struck by the irony of the minor premise of her observation. New scientific objects need not only be like quotidian objects, but sometimes they may simply be, for one reason or another, literally quotidian. If culture is a good example of an invisible planet—one not directly visible until scientifically dependable, credible tools are developed (Shapin 1995)—then surely anthropology is its powerful instrument, though one that can require somewhat regular recalibration. In conversation with the other papers collected here, I suggest that the natural category of 'flowing or standing fresh water' in the Gambier Islands may indeed require a modest but meaningful recalibration in disciplinary discourse, even as locally situated Mangarevan persons are grappling with the presence (or sometimes the absence) and significance of water on the landscape and in cultural conception.

While water has a delicious history in anthropological writings, it has played a curious role in our disciplinary practice and conception, oscillating 'between natural and cultural substance, its putative materiality masking the fact that its fluidity is a rhetorical effect of how we think about "nature" and "culture" in the first place' (Helmreich 2011: 132). Recent interest in water, including Helmreich's, seems to be motivated by an increased sense of the salience of this specific substance in nature, further contextualised by the (re-)emergence of nature itself as a fresh object of anthropological inquiry (Muru-Lanning 2007, 2009, 2010; Wagner 2010, 2013). However, the ongoing 'scientisation' of water in the social sciences, in which my work is partially embedded, is caught up in Féaux de la Croix's (2011) and Helmreich's (2011) separate observations that, in order to bring water into contemporary anthropological focus, there needs to be careful attention to (and potential skepticism about) the previous deployment of the qualities and characteristics of water in theory—for instance, as a trope of globalisation in its potent currents, flows, tides, logjams or frothy bubbles. Helmreich's (2011: 133) thoughts on the 'oceanisation' of globalisation seem to deserve particular attention by scholars of Pacific places and peoples because this trend may further displace attention to fresh water in a region so wonderfully, and sometimes so problematically, dominated by the characteristic and dynamic majesty of the ocean.

Two natural tropes appear to consistently play a role in centering Mangarevan conceptions, understandings and experiences of the threaded relationships between the past and the present. These are *tumu* and *puna*—roots and springs. The principles of rootedness and of frothy 'forth-springingness' seem to play numerous roles in both everyday conversations and the heightened talk contained in songs, chants and proverbs. The role of these highly salient metaphors can be observed in the preserved texts of the nineteenth and early twentieth centuries, as well as in contemporary talk.

To a certain extent, I suspect the centrality of these tropes in organising and articulating local understandings of individual and community relationships helps to explain some of the anxieties about lost springs and blocked waters. Leaving the *tumu* ('roots') of the present aside, the disappearance or movement of springs on a named and intimately felt landscape, the diversion of a watercourse for agricultural purposes and the cessation of riverine irrigation seem to bring into awareness a strong sense of a lost past, of a disconnection with honored antecedents, the result of which may be fundamentally challenging to senses of individual and collective self.

The exploration of the material contexts of everyday human conditions, in particular the seemingly ubiquitous facets of nature domains, has been a key disciplinary turn in recent years (Escobar 1999; Ingold 2000, 2011). Such socio-spatial objects as forests, rivers, lagoons and winds have perhaps been too frequently overlooked, pre-theoretically intuited as universal terms that are automatically translatable across cultural regimes (Steinberg 2001; Helmreich 2011). Water, in all of its characters and qualities, seems as deserving of renewed attention in Eastern Polynesia as any of the other latently under-scrutinised features in anthropological attempts to make sense of others and other natures, or to make sense of the way we make sense of others and their natures. The material contexts at the intersection of nature and culture in everyday life offer clear purchase on the core anthropological project of translating the lives of others in such a way that conversations across and between them can take place. At the same time, the emergent interest in extending the anthropological project afresh to such renewing sources as rivers and springs, or to water itself, is also transparently a comment on the legacy of disciplinary traditions, the legacies of past objects of inquiry, and gaps and lacunae in earlier anthropological works that are now called vigorously into attention by serious contemporary concerns with the state of nature, with environmental degradation, or with ecological collapse and its implications for social stability and cultural transmission.

Thus, while embracing the opportunities offered by the investigation of *puna vai* ('springs') to think about the historical figurations of landscapes, places and peoples, I also appreciate the opportunity offered by the investigation of rivers in Oceania to think with, splash around, dive in and otherwise muddy these sinuous, fluid ethnographic objects. For these are rivers that may only flow in stories, like the springs and founts that flow on paper but appear to have moved or been lost in the landscape, or waters that do in fact flow but have lost their apparently previously mythical or culturally heightened qualities.

To return to, and conclude with, thoughts introduced at the outset of this chapter, I note that the moral colouration of the stories 'we' tell about standing or flowing waters would also be familiar in a North American or European cultural context, where environmental policy and rhetoric are not always instrumental but are also often about notions of the past and of past communities. Such stories reflect complex feelings of nostalgia and loss, or of time that somehow should have been 'ours' being made other.

Acknowledgments

A flowing stream of gratitude is owed to the Pupu Vaka Varaga ko Nikola Mamatamoe, Bruno Schmidt and Daniel Teakarotu for the *varagaraga* in 2002–03, and again in 2009, that made this chapter possible. Hearty thanks also to Rai Chaze, Ena Manuireva, Lisa Ledvora, Eva-Marie Dubuisson, Robert Sullivan, Jerry Jacka and John Wagner for suggestive readings and important editorial advice. Finally, *maro'i ake ki te 'u* Mangareva for discussions, insights, *makararaga* and *'akaporoturaga. Maro'i 'aka'ou.*

References

Anderson, A., E. Conte, P.V. Kirch and M. Weisler, 2003. 'Cultural Chronology in Mangareva (Gambier Islands), French Polynesia: Evidence from Recent Radiocarbon Dating.' *Journal of the Polynesian Society* 112: 119–140.

Blust, R., 2013. *The Austronesian Languages.* Canberra: The Australian National University, Research School of Pacific and Asian Studies (Pacific Linguistics 602).

Buck, P.H., 1938a. *Vikings of the Sunrise.* New York: Frederick A. Stokes Company.

——, 1938b. *The Ethnology of Mangareva.* Honolulu: Bernice P. Bishop Museum (Bulletin 157).

Burenhult, N., 2008. 'Streams of Words: Hydrological Lexicon in Jahai.' *Language Sciences* 30: 182–199. doi.org/10.1016/j.langsci.2006.12.005

Burenhult, N. and S.C. Levinson, 2008. 'Language and Landscape: A Cross-Linguistic Perspective.' *Language Sciences* 30: 135–150. doi.org/10.1016/j.langsci.2006.12.028

Butaud, J.F., 2009. 'Les Gambier, un Archipel à la Végétation Naturelle Relicturelle et à la Flore Patrimoniale Menacée.' *Bulletin de la Société des Études Océaniennes* 315/316: 99–140.

Cablitz, G.H., 2008. 'When "What" is "Where": A Linguistic Analysis of Landscape Terms, Place Names and Body Part Terms in Marquesan (Oceanic, French Polynesia).' *Language Sciences* 30: 200–226. doi.org/10.1016/j.langsci.2006.12.004

CESCEN (Commission d'Enquête sur les Consequences des Essays Nucléaires), 2006. *Les Polynésiens et les Essays Nucléaires: Indépendance Nationale et Dépendance Polynésienne.* Papeete: Assemblée de la Polynésie Francaise.

Charpentier, J.M. and A. François, 2015. *Atlas Linguistique de la Polynésie Française / Linguistic Atlas of French Polynesia.* Berlin and Papeete: De Gruyter and Université de la Polynésie Française. doi.org/10.1515/9783110260359

Conte, E. and P.V. Kirch, 2011. 'One Thousand Years of Human Environmental Transformation in the Gambier Islands (French Polynesia).' *Terra Australis* 29: 253–264.

Cronon, W., 1992. 'A Place for Stories: Nature, History, and Narrative.' *Journal of American History* 78: 1347–1376. doi.org/10.2307/2079346

Daston, L. (ed.), 2000. *Biographies of Scientific Objects.* Chicago: University of Chicago Press.

Diamond, J., 2005. *Collapse: How Societies Choose to Fail or Succeed.* New York: Penguin.

Dunis, S., 2006. 'Seaworthy Myths: A Tentative Interpretation of 'Atua Mata Riri, a Rapa Nui Procreation Chant.' In *The Reñaca Papers: VI International Conference on Rapa Nui and the Pacific.* Los Osos (CA): Easter Island Foundation.

Dyen, I., 2005. 'Some Notes on the Proto-Austronesian Words for "Water".' *Oceanic Linguistics* 44(1): 1–11.

Escobar, A., 1999. 'After Nature: Steps to an Anti-Essentialist Political Ecology.' *Current Anthropology* 40: 1–16. doi.org/10.1086/515799

Féaux de la Croix, J., 2011. 'Moving Metaphors We Live By: Water and Flow in the Social Sciences and around Hydroelectric Dams in Kyrgyzstan.' *Central Asian Survey* 30: 487–502. doi.org/10.1080/02634937.2011.614097

Feinberg, R., 1980. 'History and Structure: A Case of Polynesian Dualism.' *Journal of Anthropological Research* 36: 361–378. doi.org/10.1086/jar.36.3.3629530

Green, R.C. and M.I. Weisler, 2000. *Mangarevan Archaeology: Interpretations Using New Data and 40 Year Old Excavations to Establish a Sequence from 1200 to 1900 A.D.* Dunedin: University of Otago (Studies in Prehistoric Anthropology 19).

——, 2002. 'The Mangarevan Sequence and Dating of the Geographic Expansion into Southeast Polynesia.' *Asian Perspectives* 41: 213–241. doi.org/10.1353/asi.2003.0006

Greenhill S.J. and R. Clark, 2011. 'POLLEX-Online: The Polynesian Lexicon Project Online.' *Oceanic Linguistics* 50: 551–559. doi.org/10.1353/ol.2011. 0014

Gregory, H.E., 1934. *Report of the Director for 1933.* Honolulu: Bernice P. Bishop Museum (Bulletin 133).

Hau'ofa, E., 1998. 'The Ocean in Us.' *Contemporary Pacific* 10: 392–410.

Helmreich, S., 2011. 'Nature/Culture/Seawater.' *American Anthropologist* 113: 132–144. doi.org/10.1111/j.1548-1433.2010.01311.x

——, 2014. 'Waves: An Anthropology of Scientific Things.' *HAU: Journal of Ethnographic Theory* 4(3): 265–284.

Ingold, T., 2000. *The Perception of the Environment: Essays on Livelihood, Dwelling and Skill.* London: Routledge. doi.org/10.4324/9780203466025

——, 2011. *Being Alive: Essays on Movement, Knowledge and Description.* London: Routledge.

ISPF (Institut de la Statistique de la Polynésie Française), 2015a. 'Affiche Recensement 2012.' Viewed 5 August 2016 at: www.ispf.pf/docs/default-source/rp2012/AFFICHE_ISPF_RECENSEMENT_2012.jpg?sfvrsn=0

——, 2015b. 'Population sans Doubles Comptes, des Subdivisions, Communes et Communes Associées de Polynésie Française, de 1971 à 1996.' Viewed 5 August 2016 at: www.ispf.pf/bases/Recensements/Historique.aspx

Janeau, V.F., 1908. *Essai de Grammaire de la Langue des Iles Gambier ou Mangaréva.* Braine-le-Comte: Zech et Fils.

Kirch, P.V., 2007. 'Three Islands and an Archipelago: Reciprocal Interactions between Humans and Island Ecosystems in Polynesia.' *Earth and Environmental Science Transactions of the Royal Society of Edinburgh* 98: 85–99. doi.org/10.1017/S1755691007000011

Kirch, P., E. Conte, W. Sharp and C. Nickelsen, 2010. 'The Onemea Site (Taravai Island, Mangareva) and the Human Colonization of Southeastern Polynesia.' *Archeology in Oceania* 45: 66–79. doi.org/10.1002/j.1834-4453.2010.tb00081.x

Kirch, P.V. and J.L. Rallu (eds), 2007. *The Growth and Collapse of Pacific Island Societies: Archaeological and Demographic Perspectives.* Honolulu: University of Hawai'i Press.

Kuhn, T., 1970. *The Structure of Scientific Revolutions* (2nd edition). Chicago: University of Chicago Press.

Laval, H., 1938. *Mangareva: L'Histoire Ancienne d'un People Polynésien*. Paris: Librairie Orientale Paul Geuther.

Levinson, S.C., 2008. 'Landscape, Seascape and the Ontology of Places on Rossel Island, Papua New Guinea.' *Language Sciences* 30: 256–290. doi.org/10.1016/j.langsci.2006.12.032

Malinowski, B., 1931. 'Culture.' In E. Seligman and A. Johnson (eds), *Encyclopedia of the Social Sciences*. New York: Macmillan.

——, 1992 [1948]. *Magic, Science and Religion and Other Essays*. Long Grove (IL): Waveland Press.

Mamatui, I., n.d.a. 'Manuscript PAC Mangareva 3.4.' Honolulu: Bernice P. Bishop Museum.

——, n.d.b. 'Atonga Mangareva.' Honolulu: Bernice P. Bishop Museum (manuscript).

Mawyer, A., 2008. 'The Oceanic Drift in Polynesian Linguistics.' *Language and Communication* 28: 363–385. doi.org/10.1016/j.langcom.2008.01.012

——, 2014. 'Disoriented Space in the Gambier, French Polynesia.' *ETHOS: Journal for the Society for Psychological Anthropology* 42: 273–301.

——, 2015. 'Wildlands, Deserted Bays, and Other "Bushy" Metaphors of Pacific Place.' In J.A. Bell, P. West and C. Filer (eds), *Tropical Forests of Oceania: Anthropological Perspectives*. Canberra: ANU Press (Asia-Pacific Environment Monograph 10).

McAnany, P.A. and N. Yoffee (eds), 2010. *Questioning Collapse: Human Resilience, Ecological Vulnerability, and the Aftermath of Empire*. Cambridge: Cambridge University Press.

Muru-Lanning, M., 2007. 'Tupuna Awa and Sustainable Resource: Knowledge Systems of the Waikato River.' *MAI Review* 1 Article 6. Viewed 6 June 2018 at: review.mai.ac.nz/MR/article/viewFile/30/30-30-1-PB.pdf

——, 2009. 'River Ownership: Inalienable Taonga and Impartible Tupuna Awa.' *Sites: A Journal of Social Anthropology & Cultural Studies* 6(2): 32–56.

——, 2010. '*Tupuna Awa* and *Te Awa Tupuna*: An Anthropological Study of Competing Discourses and Claims of Ownership to the Waikato River.' Auckland: University of Auckland (PhD diss.).

Rensch, K.H., 1991. *Tikionario 'Arani—Mangareva*. Canberra: Archipelago Press.

Sahlins, M., 1958. *Social Stratification in Polynesia*. Seattle: University of Washington Press (American Ethnological Society Monograph 29).

Shapin, S., 1995. *The Social History of Truth*. Chicago: University of Chicago Press.

Steinberg, P.E., 2001. *The Social Construction of the Ocean*. Cambridge: Cambridge University Press.

Strang, V., 2004. *The Meaning of Water*. Oxford: Berg.

Thibault, J.C. and A. Cibois, 2011. 'From Early Polynesian Settlements to Present: Bird Extinctions in the Gambier Islands.' *Pacific Science* 66(3): 1–26.

Tiripone, n.d. 'Atonga Mangareva.' Honolulu: Bernice P. Bishop Museum (manuscript).

Tregear, E., 2009. *Mangareva Dictionary: Gambier Islands* (2nd edition). Tipaerui: Société des Etudes Océaniennes.

Vallaux, F., 1994. *Mangareva et les Gambier*. Tahiti: Etablissement Territorial d'Achats Groupes.

Wagner, J., 2010. 'Water Governance Today.' *Anthropology News* 51(1): 5–9. doi.org/10.1111/j.1556-3502.2010.51105.x

—— (ed.), 2013. *The Social Life of Water*. New York: Berghahn Books.

5. Riverine Disposal of Mining Wastes in Porgera: Capitalist Resource Development and Metabolic Rifts in Papua New Guinea

JERRY K. JACKA

Introduction

It is mid-morning and Epe and I are walking along the four-wheel drive road in Tipinini that connects the hamlets of Kolarika, Lese and Yomondaka with the main road (Figure 5.1). We are on our way to his garden in Yomondaka to join his wife, two daughters and son who are at work, weeding and harvesting. As we cross the bridge over the Wateya River, we see Jonah Wuambo just upstream, sitting fully clothed underneath a small waterfall in a pool of water. The water pours directly onto Jonah's head, encasing him in a curtain of frigid water. 'Jonah must be sick', Epe says. Despite Porgera's location at only 5° south of the Equator, the high altitude means that the rivers are cool, and sitting in a cold waterfall while sick seems counter-intuitive to me—a point that I make to Epe. 'No,' he replies, 'the water will wash his sickness away. It will make him feel better later.' Epe and I stand at the railing of the bridge watching Jonah for a few minutes; he eventually notices us and waves weakly. We wave back and continue on to Epe's family's garden.

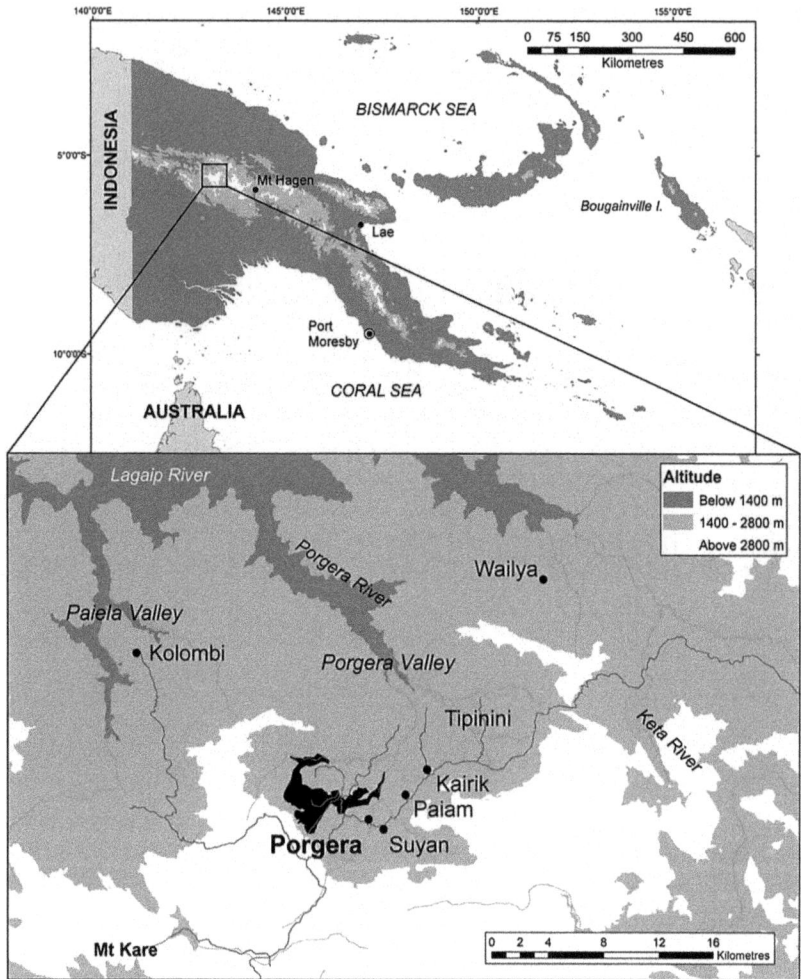

Figure 5.1 The Porgera Valley.
Source: Cartography by the author.

Water in its many forms—as rain, rivers or ponds—is an elemental force in Porgeran culture and ecology. Water, or *ipa* in both of the local languages (Ipili and Enga), figures centrally in curing rituals and in ideas about the maintenance of fertility in people, flora and fauna, and even the land. There is also an opposition between moving water and still water,[1] with water in motion considered to be more efficacious in its power to heal and nurture (see Weiner 1991: 99–105). This concept of motion highlights

---

1   See Hviding, this volume, on the opposition between salt and fresh water in a coastal community.

another key metaphor in Porgeran philosophy that contrasts movement with stasis. In most cases, stasis is a result of some form of blockage. Most illnesses are perceived as being due to the inability of fluids to move through the body. Cures for sickness either utilise or emulate the free flow of liquids, and sitting under waterfalls, drinking large quantities of water and invoking the names of certain sappy trees in healing spells are just some of the ways that fluidity is deployed.

The cultural significance of clean, flowing water for promoting health in people and the land is in direct contrast to the activities of the Porgera gold mine. Built in 1990, the Porgera mine is one of only four mines in the world that practice the riverine disposal of mining wastes (Vogt 2012). Every day, 17,000 metric tonnes of mining tailings and waste rock are dumped into two major tributaries of the Porgera River. The past 28 years of mining activities have resulted in two massive rock glaciers, each tens of millions of metric tonnes in size, moving slowly down the river valley, erasing gardens, homesteads and past sites of artisanal alluvial mining. The Porgera River often runs blood-red from the iron oxides released during the ore crushing phase of mining. Other toxic metals are potentially altering downstream ecologies and livelihoods in manifold ways (CSIRO 1996; Shearman 2001), and the ensuing perception of pollution in Porgera has profound social and cultural effects (see Beck 1992). Riverine pollution is the legacy of mining development in Papua New Guinea (PNG) (see also Banks et al. 2013). Earlier projects, such as the Panguna mine in North Solomons Province and the Ok Tedi mine in Western Province, have proven to be socially, economically and ecologically costly as downstream communities have either halted the mining project, as in the case of Panguna (May and Spriggs 1990), or been granted a multi–million dollar out-of-court settlement, as happened at Ok Tedi (Kirsch 2006, 2014).

My concern in this chapter is to examine the implications of turning a critical force in Porgeran well-being—namely rivers—into dumping grounds for resource extraction (see Figure 5.2). In a cultural context in which humans and their natural environment share life-giving powers, how does riverine pollution alter social and ecological relationships? Building upon Marx's concept of metabolism, John Foster argues that capitalism creates 'metabolic rifts' between the social and natural worlds (Foster 2000: 155–70). Or as Marx himself put it:

It is not the *unity* of living and active humanity with the natural, inorganic conditions of their metabolic exchange with nature, and hence their appropriation of nature, which requires explanation or is the result of a historic process, but rather the *separation* between these inorganic conditions of human existence and this active existence (Marx 1973: 489, emphasis in original).

For Marx, the natural world was 'the inorganic body' of humanity (Foster 2000: 72), and he conceived of the metabolic rift in terms of the way that capitalist agriculture consumed the fertility of the soil. I argue that these same processes are at play in Porgera but in a fluvial form, so my goal in this chapter is to explore the social and ecological implications of the metabolic rift that mining has created between Porgerans and their aquatic environments.

Figure 5.2 The Porgera gold mine.
Source: Photo by the author.

In the following sections, I first describe the ethnographic setting, then focus on the importance of water in Porgeran cosmology. Next I discuss the sociality of rivers, addressing how rivers are not only borders between social groups, but are also used to depict the personal networks that cognatic kinship creates. I also explore rivers as resources that provide additional benefits to a terrestrially focused subsistence horticultural base. In the final section of the chapter, I examine the impacts of riverine tailings disposal on the human and biotic communities downstream from the Porgera mine. My conclusions address the implications of the metabolic rift for Porgeran society.

The Porgera Valley

The Porgera Valley is home to Ipili-speaking subsistence horticulturalists who numbered just over 5,000 in the early 1980s. Since mining development began, outsiders—especially the more numerous Enga speakers to their east—have poured into the valley seeking job opportunities and improved livelihoods that resource development promises. Many of these migrants were welcomed onto Ipili lands, where they built houses and made gardens, eventually becoming integrated into local clans through marriage and exchange ties. With a population just over 22,000 in the 2000 census, today it is estimated that there are around 50,000 people in the valley. Conflicts over land and the scarcity of firewood and building materials are common. Monetary payments in the form of mining royalties and compensation for riverine pollution are another source of frequent outbreaks of violence. Porgeran society is bifurcated between the 'landowning' clans that receive mining benefits due to their lands being within the boundaries of the special mining lease and the much larger numbers of 'non-landowners' who suffer the costs of hosting a mine in the valley but receive little in the way of development benefits (Jacka 2001a, 2007, 2015). Aletta Biersack (2006) has already described the political struggles associated with riverine pollution between landowners, non-landowners and the Porgera Joint Venture (PJV), which operates the mine. In this chapter, I build on her analysis to examine the cosmological impacts of riverine pollution and also to explore some of the ecological and social consequences of riverine degradation.

In Porgera, it is hard to get away from the sound of rushing water. Annually, over 3,700 mm of rain falls upon the mountainous landscape, turning rushing streams into roaring, boulder-choked rivers further downstream. Geologically, Porgera is comprised of a limestone matrix interbedded with layers of mudstones and siltstones. Due to the porous nature of the limestone, several waterfalls gush from holes midway down the steep cliffs that form the valley's mountainous southern border. Elsewhere in the high-altitude limestone karst country above these mountains, entire rivers disappear into sinkholes only to emerge kilometres away from whence they vanished. Consequently, Porgerans envision the landscape as riddled with tunnels through which it is not only water but also ancestral and nature spirits that move. In fact, in order to talk about water and rivers in Porgera, it is necessary to also talk about spirits and the roles they play in regulating water and environmental productivity.

The Cosmology of Water

There is a fundamental distinction in Porgeran thought between wetness and dryness. In general, wetness and its associated qualities are considered to be a normal or natural state of things, while dryness is a sign of sickness, old age or infertility. Traditionally, collective male bachelor fertility cults and individual female maturation practices used water in order to grow young people into marriageable, fertile adults (Biersack 1998a, 2004). Water, in both of its manifestations as rain and rivers, is embedded in ideas about spirits, the need for moral behaviour in regard to spirits, and moral breaches of reciprocity as likely to result in world-ending calamities. In such a mountainous region, the destructive capacity of water is a common topic of conversation due to the numerous rain-induced landslides and the torrential rains that swell the waterways and erode adjacent gardens. Yet water is also recognised for its creative abilities to heal, promote growth and combat desiccation and dryness, which are symptoms of human illness and ecological decline. Water is also an important metaphor in that its flow is conceptually believed to ritually link together diverse cultural groups across the PNG highlands (Goldman and Ballard 1998).

Cosmologically, water has the capacity to bestow immortality. Engans (Wiessner and Tumu 1998), Hulis (Ballard 1998; Wardlow 2006) and Porgerans all recite a story about a missed opportunity for immortality when the first baby was born. Piyapa Inguni of Wailya recited the story as follows:

> There is a mountain near Mulitaka that has no trees on it; it is only covered with grass. It is called Mt Mungalo. Long ago, there was a woman on Mt Mungalo and she had just given birth. Her husband told her that he was going to go and get water for the baby from a sacred lake [*ipa tawe*]. He told his wife that if the baby cried she couldn't give it breast milk; she had to wait until he returned with the water. While the man was gone, the baby cried and cried and cried. Finally, the woman let the baby drink from her breast. Then from afar the man cried out, *Atambiape?* ['Live?']. He heard nothing, so he yelled out, *Umambiape?* ['Die?']. The woman

responded affirmatively, *Uuuu*. When she responded to *umambiape*, he dumped the sacred water on the ground. *Atambiape* is the same message that is inside the Bible or that the government tells us; it means we can't fight or get cross. *Umambiape* has to do with fighting and killing, and we followed this path. If the baby had drunk the water, we would live forever. Since it drank the milk, we are all mortal.

Water, since it comes from the sky people,[2] and as a one-time potential source of immortality, has the power to wash away sickness and restore health. As described at the beginning of this chapter, Jonah's immersion into the cold waterfall reflects this belief in the curative powers of water. Water is not just important for humans, but figures critically in key ideas about Porgeran ecology. A key principle in Porgeran ecological understanding centres around the idea that human and spiritual activities and environmental processes facilitate the movement of a vital life force called *ipane* (literally 'grease') through people, plants, animals and the land. The word *ipane* is comprised of two morphemes, *ipa* ('water') and *ne* ('like'), which suggests that grease is something 'like water'. Just as rivers flow across the landscape, disappear into sinkholes and move through the land to reappear elsewhere, *ipane* flows through the land into plants, into humans and animals that consume those plants, and then is released back into the land when animals are sacrificed and people die. One of the main purposes of *ipane* is to replenish people, plants and land that are perceived to be otherwise dry.

In several publications, Aletta Biersack (1995, 1998b, 2005, 2011) has argued that human life for the Ipili is governed by a 'sacrificial principle'. What she means by this is that the maintenance and reproduction of human life depletes a person's life force. Gardening, hunting, childbirth, economic exchange, rearing children and so forth exhaust a person to the point that death is inevitable. While most of us are aware of this at a subconscious level, for Ipili speakers it is a central preoccupation. Children are literally said to 'take the place' (*panda mia*) of their parents (Biersack 2011: 230). The principle of a life for a life is also central to healing rituals in which pigs are killed for sick people, and to the logic of warfare and revenge killings whereby deaths are compensated through payments of pigs.

---

2   *Ipa tawe* literally means 'sky water', and refers to small high-altitude lakes associated with the 'sky people'.

The signs that someone has spent their life in toil under the sacrificial principle are manifested in a person's outward appearance. Wrinkles, white hair and loss of vitality are all indicators of ageing, but also indicate that people have depleted their *ipane* through the work of production and reproduction. Two vital life fluids, breast milk and semen, contain *ipane*: breast milk is called *andu ipane* ('breast grease'), while semen is called *ui ipane* ('penis grease'). By producing children, parents gradually diminish their own life force. Agricultural activities also involve the flow and transfer of grease, since all land is considered to contain varying amounts of *yu ipane* ('ground grease'). Over time, as the land is cultivated, a garden eventually loses its grease, resulting in a new garden being planted where there is more grease. Over even longer periods of time, there is a general consensus that the grease has been exhausted from the entire region, and ritual activities of varying intensity, involving one or more social groups, are required to revitalise the world.

Pools of still (immobile and unhealthy) water are thought to contain various spirits that affect Porgerans. Every clan has a sacred pool inhabited by the spirits of deceased ancestors in the high-altitude rainforest above the hamlets. The pools are called *ipa ne*,[3] literally meaning 'the water that eats'. Traditionally, illnesses were (and sometimes still are) believed to be caused by deceased clan ancestors. Various rituals were performed to feed the ancestors, by offering them 'food' in the form of the aromas of cooked meat or the blood of pigs poured into the ground, or by the actual delivery of meat to the pools themselves. In the case of the latter, multiple informants describe the water in the pool rising up to snatch the bundle of meat off a long pole that ritual supplicants held over the water. Today, attitudes about these pools are varied. I have previously described a long-abandoned ceremony called a *kolo* that was held in 1999 by the Paiam clan on the site of the Paiam supermarket (Jacka 2001b). The building of the supermarket had destroyed the Paiam clan's sacred pool, and when food and drinks were mysteriously disappearing from the supermarket, the Paiam held a *kolo* to placate their angered ancestors. On another occasion, after a group of young men took me to see the Pakoa clan's sacred pool, I awoke in one of their houses the following day with my first-ever migraine headache, which everyone attributed to us having disturbed the clan's ancestors. Nevertheless, a few months later, I accompanied my

---

3   Not to be confused with *ipane* ('grease') since the words are pronounced with different stress on the syllables: I-pa NE versus i-PA-ne.

research assistant Ben and his father Pes to the site of the Tokoyela clan's sacred pool. It had long since silted over, and Pes nonchalantly walked all over the grassy surface of the former pool, chopping at the grass with his bush knife while telling me about how afraid they used to be of this area because of the ancestral spirits (see Figure 5.3).

Figure 5.3 Pes Bope at the Tokoyela *ipa ne*.
Source: Photo by the author.

Other pools in the high rainforest were also held to contain powerful forces. These pools were associated with sky women (*tawe wanda*) who had a mythical power to transform boys into powerful men. In all of the myths, a small good-for-nothing younger brother is shunned by his older brothers. After being cast out of the house, he wanders the high rainforest until he runs into a beautiful woman. She takes the young boy and coaxes him under the water. When he re-emerges, he has been transformed into a tall, muscle-bound warrior. The sky woman then plies him with pigs, shell wealth and resplendent bodily ornamentations like bird of paradise headdresses, marsupial fur headbands and woven cane armbands. Thus transformed, the younger brother returns to his hamlet, where he is desired by all of the young women and envied by his brothers. A number of male puberty ceremonies among the Ipili and Enga attempted to ritually unite participants with a sky woman in order to bring about these positive transformations encoded in myth (Meggitt 1964; Biersack 1998a, 2004; Wiessner and Tumu 1998).

Tawe wanda also play beneficial roles in human affairs. Biersack's (1982, 1998a, 2001, 2004) writings on the beautification of young people of both sexes show that the sky women are largely responsible for 'growing' adolescent males into marriageable and sexually active adults. Growth in general is thought to occur through female agency. Pregnancy also associates women with bodily growth. Additionally, while men cut and fence gardens, women plant and tend the crops and thus 'grow' the gardens. Growth does not occur naturally, but 'transactionally, as the beneficiary of someone's action' (Biersack 1982: 241, emphasis removed). Through proper observance of menstrual taboos, wives help grow their husbands' skin by depositing their menstrual wastes at the base of *tato* (southern beech or *Nothofagus*) trees, which are the only trees considered strong enough to withstand the powerful influences of menstrual blood. During puberty rituals, young bachelors 'married' *tawe wanda* who were then responsible for 'growing' them into marriageable adults. While there were no equivalent rituals for young women, they did go out into secluded spots in the residential area in the mornings and shake dew onto their skin, which was said to grow their skin, breasts and hair in anticipation of marriage and motherhood.[4] Dew, rain and lightning are all associated with sky people. Among the Enga, sky women are said to cause rain, sprinkling water from a lake located in the sky (Talyaga 1982: 67).

4　See Biersack (1998b) for a full description of young women's growth rituals.

While water has the power to heal sickness and cause maturation, there are also malevolent forces associated with rivers and pools. Trickster spirits called *ipa titi* ('water fools') live in the rivers (see Goldman 1998). Generally, these spirits are merely mischievous, and will hide people's axes or spades while they are resting from gardening work, but occasionally they are more dangerous. When women are planting sweet potatoes, they hang their infants in net bags from tree limbs next to where they are working to keep them close but out of harm's way. If this is done too close to a stream or river, *ipa titi* will sometimes swap their own babies with the human babies. Children that are physically deformed or mentally challenged are said to be the products of such transfers. As a result, Porgerans prefer not to make gardens adjacent to streams and rivers, which has the added ecological benefit of reducing erosion and sedimentation in the rivers.

As seen in this and the previous section, water serves as a critical feature through which to observe the interface of human and spiritual worlds in Porgera. Water is the medium through which humans interact with spirits, and water as both precipitation and in the flow of rivers provides evidence of spiritual forces acting upon the land. In the next section of this chapter, I turn away from spirits and the cosmology of water to examine some of the ways that water defines sociality in the area.

The Sociality of Rivers

Rivers, like mountain ranges, are borders that separate social and ethnolinguistic groups in PNG (Weiner 1991; Feld 1996; also Wagner, this volume), while major streams often function as boundaries between clans. The Wateya River, introduced at the beginning of the chapter, is the border between the Tokoyela and Yomondaka clans; however, in songs and oral literature, Wateya is a metaphorical way of referring to someone from either of these clans. The Wateya River is both a symbol of the separateness of two different social groups and a symbol of their shared involvement with that waterway. Rivers, therefore, both separate and unify social groups (see Feld 1996).

The first funeral I attended in Porgera was of a woman, Tapiama, who died unexpectedly in her middle to late 40s. She was one of Samson Kaiyama's 'aunties', and he came by my house early one morning to see if I wanted to attend her funeral with him. I accompanied Samson, absolutely uncertain of what to expect. We walked for about an hour along a rough track through forests and gardens to Lese, the main hamlet of the Pakoa clan. As we climbed up to the cleared space in the centre of the hamlet in which all social functions take place, we could hear the wailing and crying of mourners. Samson burst into a rhythmic wailing that matched that of the other mourners, and went and joined the group of people huddled around Tapiama's open casket. I joined the crowd of several dozen onlookers who conventionally sit in the grass around the main mourners and support them by their presence at the funeral.

After 15 minutes or so, Samson joined me. Samson was one of my closest neighbours in Kolatika, as his sub-clan, the Tokoyela-Pinawi, had donated the land my house was built on to the district authorities for development purposes. A few minutes after Samson sat down, Tapiama's oldest daughter, a woman in her 20s, began to wail and sing a song while holding her mother's head between her hands. Samson urged me to record the song and said that we would talk about it later. Later that afternoon, back in my house, Samson and I listened to the recording. This is a translation of the song:

> Tapiama,
> Woman of the Tupa River,
> Woman of the Lombopi River,
> Woman of the Sangga River,
> You've gone too soon.
>
> Tapiama,
> Woman of Mt Kalepe,
> Woman of Mt Paiye,
> Woman of Mt Auwakome,
> You've gone too soon.

Tapiama's daughter repeated these verses over and over, occasionally adding a new river or mountain to her mourning song. Samson explained the significance of these place names to me, noting that the Tupa, Lombopi and Sangga rivers were boundaries between the Pakoa clan and the Nomali, Yawanakali and Lake clans respectively. Tapiama's funeral had been attended by Nomali, Yawanakali and Lake people with whom she shared kinship relations. Mt Kalepe was a metonym used specifically to refer to the Tokoyela clan whose members were also Tapiama's kin. Paiye and Auwakome are the mountains of the Pakoa people. In essence, what Tapiama's daughter's mourning song had signified was that, despite clans in Porgera being associated with particular landholdings, people's paths through life, both through their everyday practices and through their kinship linkages, are defined by the rivers and mountains that both separate and unify social groups. Her mourning song wove together the places and people that were instrumental in Tapiama's identity as a social being.

Major rivers are also used in the Porgera area to express oppositional identities (see Silverman, this volume). Porgerans define themselves in opposition to the Hewa people, living north of the Lagaip River, whom they characterise as demons and cannibals in stories. In one story about the origins of agriculture, Tisapa Yawe tells of the first woman 'who had no parents or kin and was the mother of all Porgerans' and only ate wild plants and the leaves of trees. She lived at the mouth of the Porgera River, and one night a torrential rainstorm flooded the river. In the morning, she found all of the domesticated plants that Porgerans eat today—sweet potatoes, taro, greens—and an axe. She tasted the foods and found them delicious so she cut a garden out of the forest. Afterwards, she planted pandanus trees all along the south banks of the Lagaip River to mark the border between her place and the land of the Hewa.

Another example of rivers expressing oppositional identities comes from a myth about an ancient battle between the Keta River and the Lagaip River, in which the Lagaip pushed the Keta into its present-day watershed (see Figure 5.1). The Keta River is the homeland of a number of Western Enga clans who are traditional enemies of Eastern Porgeran clans. In 1988, a tribal fight broke out between two Eastern Porgeran clans. The Undiki called upon some of their Western Engan allies to come and support them in the fight, which they did. The Tokoyela tried to find some allies in Porgera to assist them, but as this was a period of time in which a gold rush was occurring at nearby Mt Kare, and mining development was already underway at Porgera, few clans were willing to get involved in the conflict. However, later that year, a Western Engan man who was an ally of the Undiki was heard boasting that the Keta River was going to wipe out the Porgera River in this tribal fight. Invoking this myth in this manner, and reinterpreting it to represent contemporary ethnic opposition between Porgerans and Engans, upset several of the Eastern Porgeran clans who had remained neutral. As a consequence, they joined the Tokoyela and defeated the Undiki, driving most of them into the Western Enga area and out of the Porgera Valley.

Rivers as Resources

If rivers are important cultural, historical, symbolic and social markers in this area of the PNG highlands, what about their practical and everyday uses?

I never saw a Porgeran drink from a river. People are unsure of what others upstream may have done to or in the water. When people slaughter pigs, women take the intestines to the river and wash them out in the water; people bathe and wash their clothes in the rivers; some households use them as toilets and, as a consequence, people are unwilling to drink river water (see Figure 5.4). Springs serve as the sources of drinking water, and scattered throughout the valley are cliff faces where someone has shoved a pandanus leaf or bamboo tube into a spring, allowing a trickle of water to be directed into drinking gourds and plastic bottles.

Figure 5.4 Bathing in the Wateya River.
Source: Photo by the author.

Traditionally, there were few aquatic resources that people gathered. The main ones were frogs and eels. Pes Bope recalls the times that his mother would gather frogs:

I would often go out and gather *pitpit* (*Miscanthus floridulus*) for my mother to make torches so that she could catch frogs. Catching frogs and birds at night was women's work. When I went out in the day to get *pitpit*, no one could see me. If they saw me they would know what was going on and send *yama au* [a type of illness]. So I hid and filled the house full of dried *pitpit*. My mother would come home from the garden and be really happy because she could go frog catching. In the afternoon she would take the *pitpit* down to the edge of the Tupa River. While waiting there for the sun to go down she would cook some food and eat. She would start bundling the *pitpit* together for torches. Once the sun went down and it got dark, she would light a torch and start catching frogs. They would be on the tree branches, rocks, next to the river, on the ground. She'd catch *kambupara*, *wa*, *kiyakonge*, *angimonge*, *doyane*, *karakara*, *pandamonge*, and *kapamu* [all different frog species]. Sometimes snakes too. She'd be out there all night catching frogs. Just before sunrise she'd come home. Whenever she knocked at the door I was always so excited. I'd ask her what she caught. She always said, 'Oh, I didn't get anything.' I'd hug her and ask, 'Mom, Mom, did you get any frogs?' She'd come in and her *bilum* [net bag] would be close to bursting with frogs. She'd take the *bilum* down and distribute frogs to everyone in the house. We'd take our share and wrap it in ferns and then we'd take it to the *pombata anda*. This means 'house of the grubs'. Anything that had meat—grubs, pig, frogs, marsupials, birds—we'd *mumu* [cook in an earth oven] in these *pombata anda* houses.

Pes's narrative highlights the importance of meat in local diets. The illness called *yama au* is very similar to another illness called *uyenda*. Whereas *uyenda* requires someone to see you eating something, *yama au* can be sent by someone merely thinking of the fact that you are getting ready to eat meat. *Yama au* is also much deadlier than *uyenda*, and hunters use magic to block *yama au* that people from the hamlets send to them while thinking about what the hunters are eating in the forest. Loosely translated, *yama au* means something like 'spirit carried', and since the illness is caused by a spirit (*yama*) it is more difficult to treat than *uyenda*, which is physiological.

Prior to the mid-twentieth century, fish (other than eels) were not part of the Porgeran diet. Rainbow trout (*Oncorhynchus mykiss*) were introduced into various highland waterways in the mid-twentieth century, both by private parties and the Australian colonial government. By 1992, there was a stock of 10,000 fingerlings in the reservoir of the Waile Creek dam in Porgera, and rainbow trout are now in all of the main waterways in the area. Trout fishing in Porgera is currently a cross between a pastime and a subsistence activity undertaken by boys and young men.[5]

Today, few people rely on the rivers for any subsistence activities. I am uncertain whether this is due to the exotic trout and/or mining wastes decimating the population of native frogs and eels in the Porgera River, or whether they are no longer gathered because of the abundance of tinned fish available in the trade stores. This of course does not mean that the rivers are no longer important for Porgerans. As detailed thus far in this chapter, water and watercourses are central to Porgeran ideas about ecology, cosmology and sociality. And there was one resource that came from the rivers that was essential to Porgeran livelihoods prior to the era of large-scale mining—the alluvial gold whose deposits came to be called 'second gardens' by most Porgerans.

These deposits were discovered in 1939 and began to be worked in the late 1940s by a few expatriate prospectors who paid Porgeran labourers with shells and steel tools. By the late 1950s, Porgerans had started mining for themselves, and in 1960 the first Indigenous claims were staked (Robinson 1960). Due to cognatic kinship reckoning, the majority of Porgerans were able to demonstrate some sort of affiliation to the claim owners or their clan, allowing them land rights and de facto rights to alluvial gold beds. In the alluvial mining era, before 1990, researchers estimated that as many as 75 per cent of all able-bodied males in the valley spent some time working, or at least had access to, the alluvial gold beds (Handley and Henry 1987).

With the demise of the global gold standard and the subsequent increase in the price of gold, outside interests—transnational mining companies and the PNG state—started seeking new ways to discover and develop Porgera's subterranean gold deposits. By 1987, the PJV had been formed and agreements were signed with the seven 'landowning' clans in regards to royalty payments, compensation for lost land, relocation houses and other development perks (Banks 1999; Filer 1999). During the negotiations for mining development, the PJV had effectively argued against the feasibility

---

5   A couple of times in 1999 and 2000, I joined Cowboy Kuraia, a young Porgeran man passionate about catching trout, and some others to fish the high headwaters of the Wateya River above Kolatika. We would creep stealthily up to the water with fishing line and hooks baited with sweet potato. Try as we might, none of us, with the exception of Cowboy, caught any fish. Graciously, Cowboy usually let me take some of his fish home to eat as a reward for my efforts.

of a tailings dam, claiming that the 'abnormally' heavy rainfall and unstable geology would lead to a massive tailings dam failure (Shearman 2001). Instead, the PJV convinced the PNG Government to allow the riverine disposal of tailings and to achieve compliance with water quality standards at a monitoring station 165 km downstream of the mine![6] The clans in Porgera immediately downstream, who had the richest alluvial gold beds, were acutely worried about their ability to continue working in what they had long called their 'second gardens' (Winchomba 1970). Moreover, these clans were not part of the group of landowning clans whose members were going to become the new beneficiaries of hard-rock mining. As we shall see, they had every right to be worried.

At this point, I want to take a step back and examine how Porgerans have conceptualised gold as a resource. Significantly, prior to its discovery by Australian patrol officers, they had no use nor word for gold, nor as far as I can tell even any awareness of it. Once its value was realised, however, gold quickly became imbued with a cosmological interpretation. This interpretation was reliant upon the idea of python spirits who moved through and lived in the tunnels believed to underlie the land. Gold is said to be their shed skins. These python spirits also played critical roles in the sacrificial rituals that Porgerans conducted to maintain the fertility of both land and people.[7] Kupiane, the python spirit associated with the Porgera gold mine, lived in the heart of what used to be Mt Watukati, which is now the centre of the mine pit. Sacrifices to Kupiane by the Tiyeni clan in the past are today interpreted by clan members as initial exchange offerings that Kupiane is now paying back to the Tiyeni, the largest landowning clan, in the form of mining wealth. That numerous other clans in the region also sacrificed to python spirits is strong evidence that the area is rife with as yet undiscovered gold deposits. Some people have also claimed that there is but one python to which everyone was making sacrifices and, by this logic, the mining wealth should be shared much more widely than it currently is.

---

6   The water quality standards were based on an Environmental Management and Monitoring Program agreed upon by the PNG Government and the PJV. The point of compliance assessment was the third of nine monitoring stations located along the Porgera, Lagaip, Strickland and Fly Rivers and in the Fly River delta.

7   As one man said to me, some people still offer sacrifices to a cave opening while on their way en route to mine for gold at Mt Kare in order to appease the python spirit, Taiyundika, 'just in case'.

The concentration of wealth among the landowning clans is particularly problematic for the downstream non-landowning clans whose members accrued most of the wealth in the alluvial mining era, especially since development of the large-scale mine has buried the alluvial beds under tons of waste rock and sediments. However, in the mid-1990s, clans that held land alongside the former alluvial beds were compensated for the loss of their income from gold. A one-time lump sum payment was made that has been followed with quarterly payments based upon the actual tonnage of waste and tailings that is put into the Porgera River. Community affairs officers insisted to me that the downstream clans will actually have earned more money this way than if they had been allowed to continue with alluvial mining. This may be so, but what cash payments have done in the hard-rock mining era is to make it easier for those people who receive royalty checks to restrict the network of individuals with whom they share their wealth. More money is coming into the valley, but it is being shared among fewer people—a point that generates considerable social tension between the haves and the have-nots. Water also plays into the way that money from mining is expected to flow. The anthropologist John Burton (2014: 46) recounts a meeting he had with some of the disenfranchised Porgerans where the lack of mining benefits was compared to 'water flowing down a tap, where a person could reasonably expect water to appear. Except … nothing came out of the tap'.

For the disenfranchised Porgerans living near the former ritual site where pigs were sacrificed to Lemeane in order to maintain the water balance of the land and ensure human and biotic fertility, the ritual site and its linkages have assumed new salience with large-scale mining development and the loss of alluvial mining opportunities. Many of the Porgerans living around the ritual site were quite emphatic that their ritual activities were part of a chain of rituals that linked Lemeane to Kupiane and to ritual sites among the neighbouring Huli and Enga (see Biersack 1998b). Huli speakers living near the ritual site of Tuandaka (Wiessner and Tumu 1998) also claim this interconnection of former ritual activities, and even argued to me that the python at Tuandaka was responsible for the oil and gas fields in the southern part of the Papuan Plateau, now the site of PNG's Liquified Natural Gas Project, as the python's urine turned into petroleum, thus enriching the landowners associated with its development. Of course, the challenge for non-landowners living near these projects is to have their claims recognised by the landowners and be able to share in the resultant resource wealth. In Porgera, thus far, this has not happened and, as a consequence, tribal fighting has increased as non-landowning clans try to extort payments from landowning clans through threats, targeted killings and the burning of government buildings located within the land boundaries of the landowning clans.

The Impacts of Riverine Pollution

The 165 km stretch of river from the Porgera mine to the water quality compliance monitoring station is called 'the mixing zone' by the PJV. This mixing zone comprises the entire Porgera River downstream of the mine, the entire lower portion of the Lagaip River from where the Porgera empties into it and the upper portion of the Strickland River. At the confluence of the Strickland and the Fly River, Porgera's mining wastes intermingle with those from the Ok Tedi mine (Kirsch 2006, 2014), and some of the material eventually flows into the Gulf of Papua. Mining companies in PNG have successfully argued for riverine disposal of mining wastes and tailings due to the high sediment loads that naturally occur in PNG's rivers, as well as to the risk that tailings dams will fail. The added inputs that come from mining, they claim, do not significantly alter the pre-mining sediment profiles. A better term for the 'mixing zone' might be 'the sacrifice zone' (Kuletz 1998), given the concentrations of toxic and heavy metals that mining releases, which are not part of naturally occurring sedimentation processes in the country's rivers. Moreover, as Stuart Kirsch (2006) has documented for people living downstream of the Ok Tedi mine, the build-up of mining sediments has caused metal-laden waters to kill riverside forests and sago gardens along vast stretches of the river. These riverine mining sacrifice zones impact the livelihoods of thousands, if not tens of thousands, of Papua New Guineans who live and garden along their banks and utilise aquatic resources in their diets.

Following an audit of PJV's waste disposal practices by the (Australian) Commonwealth Scientific and Industrial Research Organisation (CSIRO 1996), conducted in response to a report by environmental science student Phil Shearman (1995), CSIRO recommended that the PJV move their compliance assessment point to the second monitoring station approximately 40 km downstream from the mine (see Figure 5.5). This was never done, and the third station (SG3) is still the point at which compliance is assessed. Annual reports from SG3 were not published during the years that Placer Dome was the operating partner in the PJV.

In a classic case of 'green-washing', in lieu of actual data, Placer published an annual series of 'sustainability' brochures—glossy, magazine-like documents that demonstrated all of the good corporate social practices that the PJV was engaged in among the affected communities (see Burton 2014). After the acquisition of Placer Dome by Barrick Gold in 2006, the sustainability brochures were replaced with detailed reports, several hundred pages in length, illustrating the levels of heavy metals and other measures from the various monitoring stations, and rife with charts, statistics and other details beyond the comprehension of most rural Papua New Guineans. Whereas Placer Dome's technique of green-washing glossed over the impacts of riverine pollution, Barrick's version of transparency buries villagers in paper piles of mining waste documentation in a metaphorical analogue of the mining waste piles burying their lands and rivers. Both corporate strategies fail to provide accessible information to communities impacted by mining activities.

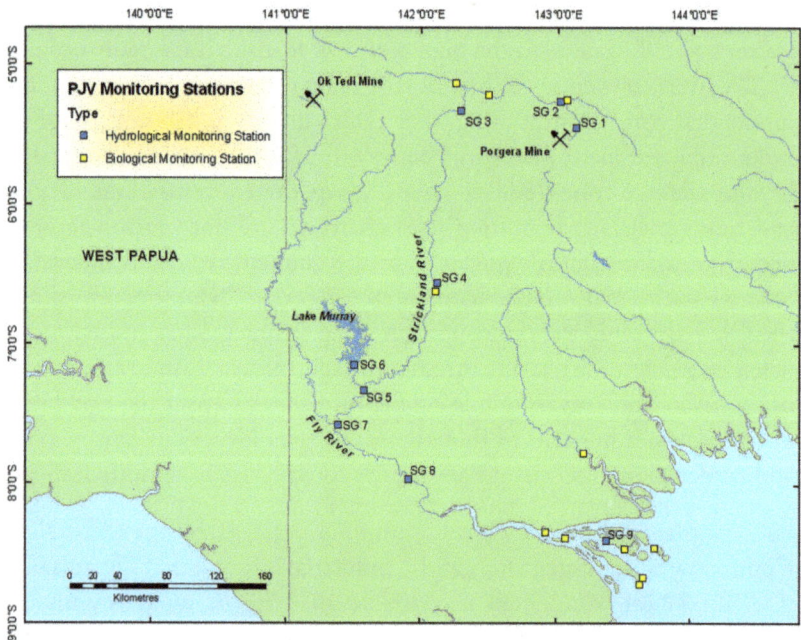

Figure 5.5 PJV monitoring stations.
Note: SG3 is 165 km downstream from the mine.
Source: Created by the author.

In 1998, following receipt of the 1996 CSIRO report, the PJV instituted what was meant to be an independent advisory body, called the Porgera Environmental Advisory Komiti ('Committee') or PEAK, to monitor the mine's impact on the physical and social environment (see Burton 2014).[8] In 2008, the PEAK decided to start issuing annual report cards on the conditions of the various river systems impacted by the Porgera mine. The data were divided into five groups of indicators: dissolved metals; metals in sediments; other water quality parameters, such as conductivity, total suspended solids, pH and cyanide; metals in fish tissues; and fish biomass. Each of these divisions was then scored as 'green—below level of concern'; 'yellow—moderate level of concern'; and 'amber—high level of concern'. The 2009 report card for the upper river at SG1 in the Porgera Valley had high levels of concern for three of the five indicators: dissolved metals; metals in sediment; and other water quality parameters. There were no data for the other two indicators. SG3, the compliance assessment point, was below level of concern for dissolved metals, had a moderate level of concern for metals in sediments and other water quality parameters, and no data for the fish indicators.

Over the years, I have received reports from local people that pigs, wildlife and humans had died after coming into contact with the water in the Porgera River (see also Biersack 2006). Penny Johnson's (2011) research on women in Porgera found that women from the villages adjacent to the mine were not only afraid to use river water for fear of the pollutants believed to be in it, but also referred to rain as '"acid rain" and the root of many illnesses in the community' (Johnson 2011: 29). Not only do people worry about the pollutants in the river; the colour of the water is also alarming from a cultural perspective. Dissolved iron oxides that result from the processing and disposal of the mining wastes have turned the Porgera River bright red on most days (see Figure 5.6). For Porgerans, 'the

8   PEAK's 'independence' from the PJV was contentious throughout its duration. In 2001, its chair Yati Bun resigned, writing: 'My conscience cannot tolerate being involved any longer with the PEAK process of expediting the continuation of riverine discharge, as when the history of Porgera is written I do not wish to be the one that oversaw Porgera's impacts and did nothing. My resolution to leave PEAK was further galvanized when I saw how my name and involvement in PEAK [was] being used in propaganda materials ...' (PEAK Minutes, October 2001). Also starting in 2001, meetings were moved from Porgera to PJV's corporate headquarters in Port Moresby. In meeting minutes from 2004, committee members worried that PEAK did not appear to be at 'arms length' from PJV, noting that 'an overly demanding watchdog might lead to PJV reluctance to provide information' (PEAK Minutes, October 2004). Following a series of investigations of human rights violations at Porgera in the early 2010s, the PEAK website was taken down and the organisation was dissolved (Philip Gibbs, former committee member, personal communication, 26 August 2017).

colour red carries a heavy symbolic load' linking the river to 'menstrual blood, which men find threatening and women find disgusting', making the river 'noxious to local sensibilities' (Biersack 2006: 233–4). Men fear menstrual blood for its abilities to 'block' a man's urethra if he inadvertently eats food that a menstruating woman has prepared or if he has intercourse with a menstruating woman. If not cured, men are said to die from this affliction. What does it mean then for a river to be perceived as a menstrual flow?

Figure 5.6 The Porgera River flowing down the 'glacier' of waste rock.
Source: Photo by the author.

Recall that the very fertility of the earth depends on the maintenance of 'grease' (*ipane*) in the land. Production and reproduction—garden labour, sexual intercourse, breast feeding, menstruation—all deplete the *ipane* in human bodies. People say that mining depletes the *ipane* in the land. A turn of phrase that Porgerans frequently use to comment on the impacts of mining is to say that 'the land is ending' (*yu koyo peya*). The gaping maw of the open pit mine eating away the land—the very resource base of Porgera's horticultural society—and the blood-red flow of the polluted Porgera River seeping out of the open gash in the land are visceral reminders to Porgerans of the costs and consequences of hosting a world-class mine on their lands.

The Porgera River thus serves as a blood-red reminder of the 'metabolic rift' and the ways that capitalist expansion tears asunder local livelihoods and ecologies. Marx (1990) used the German word *stoffwechsel* ('material exchange') to describe the metabolic processes that occur between humans and nature. What Marx was referring to was the environment as the source of food, clothing and shelter for humans, and humans as caretakers of the environment to ensure the continuing production and reproduction of both people and nature. According to Marx, large-scale capitalist production creates an irreparable rift between humans and nature because it 'disturbs the metabolic interaction between [humans] and the earth' by 'robbing' the earth of 'its constituent elements' (Foster 2000: 155–6). Not only does it rob the earth of its constituent elements; it robs the very people whose livelihoods are dependent upon the intricate social, mythological and material relations they have fostered with their environments.

Conclusion

My goal in this chapter has been to consider mining development impacts on the Porgera River in a holistic sense from the perspectives of Porgerans. When Porgerans discuss the ruination of the river, there is more at stake than just the loss of fish and frogs and the fouling of the water. The very riverbeds were dug out by the actions of ancestral hero figures that were also responsible for controlling the water balance in the land and the fertility therein. Water itself is fundamental to Porgeran ideas of cosmology and well-being. For close to 50 years, the alluvial gold resources were the second gardens that provided livelihood benefits and cash incomes in a society that was just then being integrated into the global economic system. Foreign investment and loss of control over the gold resources resonate throughout Porgeran society today in the form of often violent political clashes over the restricted benefit streams available to a minority of people in the valley.

The island of New Guinea is the only place on the planet where multinational mining companies practice riverine tailings disposal (Vogt 2012). The Porgera, Ok Tedi and Tolukuma mines in PNG, and the Grasberg mine in West Papua, all dispose of tailings that detrimentally impact the riverine environments downstream from their discharge points. In all of these cases, the mining companies were successfully able to argue that the unique geological and climatic conditions of New Guinea made riverine tailings disposal the only feasible method for tailings treatment. Given these facts, this begs the obvious question: Should mining be conducted in its current form on the island of New Guinea? The island is one of what Conservation International designates as the planet's five 'high biodiversity wilderness areas' (Mittermeier et al. 2003). It is also home to approximately one-sixth of the world's linguistic and cultural diversity. The ever-expanding growth of capitalist resource extraction promises to acutely degrade this incredible cultural and biological diversity.

Acknowledgments

Funding was provided by the Wenner-Gren Foundation for Anthropological Research and the National Science Foundation. Research permission was granted by PNG's National Research Institute. As always, my greatest appreciation goes to the people of the Porgera Valley. Ben Penale, Peter Muyu and Epe Des have helped my research in innumerable ways and I am eternally grateful for all they have done for me.

References

Ballard, C., 1998. 'The Sun by Night: Huli Moral Topography and Myths of a Time of Darkness.' In L.R. Goldman and C. Ballard (eds), *Fluid Ontologies: Myth, Ritual and Philosophy in the Highlands of Papua New Guinea.* Westport (CT): Bergin & Garvey.

Banks, G., 1999. 'The Economic Impact of the Mine.' In C. Filer (ed.), *Dilemmas of Development: The Social and Economic Impact of the Porgera Gold Mine, 1989–1994.* Canberra: Asia Pacific Press.

Banks, G., D. Kuir-Ayius, D. Kombako and B. Sagir, 2013. 'Conceptualizing Mining Impacts, Livelihoods and Corporate Community Development in Melanesia.' *Community Development Journal* 48: 484–500. doi.org/10.1093/cdj/bst025

Beck, U., 1992. *Risk Society: Towards a New Modernity* (transl. M. Ritter). London: Sage Publications.

Biersack, A., 1982. 'Ginger Gardens for the Ginger Woman: Rites and Passages in a Melanesian Society.' *Man* (NS) 17: 239–258. doi.org/10.2307/2801811

——, 1995. 'Heterosexual Meanings: Society, Economy, and Gender among Ipilis.' In A. Biersack (ed.), *Papuan Borderlands: Huli, Duna, and Ipili Perspectives on the Papua New Guinea Highlands*. Ann Arbor: University of Michigan Press.

——, 1998a. 'Horticulture and Hierarchy: The Youthful Beautification of the Body in the Paiela and Porgera Valleys.' In G. Herdt and S. Leavitt (eds), *Adolescence in Pacific Island Societies*. Pittsburgh: University of Pittsburgh Press.

——, 1998b. 'Sacrifice and Regeneration among Ipilis: The View from Tipinini.' In L.R. Goldman and C. Ballard (eds), *Fluid Ontologies: Myth, Ritual and Philosophy in the Highlands of Papua New Guinea*. Westport (CT): Bergin & Garvey.

——, 2001. 'Reproducing Inequality: The Gender Politics of Male Cults in Melanesia and Amazonia.' In T. Gregor and D. Tuzin (eds), *Gender in Amazonia and Melanesia: An Exploration of the Comparative Method*. Berkeley: University of California Press. doi.org/10.1525/california/9780520228511.001.0001

——, 2004. 'The Bachelors and Their Spirit Wife: Interpreting the *Omatisia* Ritual of Porgera and Paiela.' In P. Bonnemere (ed.), *Women as Unseen Characters: Male Ritual in Papua New Guinea*. Philadelphia: University of Pennsylvania Press.

, 2005. 'On the Life and Times of the Ipili Imagination.' In J. Robbins and H. Wardlow (eds), *The Making of Global and Local Modernities in Melanesia: Humiliation, Transformation and the Nature of Cultural Change*. Burlington (VT): Ashgate.

——, 2006. 'Red River, Green War: The Politics of Nature along the Porgera River.' In A. Biersack and J.B. Greenberg (eds), *Reimagining Political Ecology*. Durham (NC): Duke University Press. doi.org/10.1215/9780822388142-009

——, 2011. 'The Sun and the Shakers, Again: Enga, Ipili, and Somaip Perspectives on the Cult of Ain, Part Two.' *Oceania* 81: 225–243.

Burton, J., 2014. 'Agency and the "Avatar" Narrative at the Porgera Gold Mine, Papua New Guinea.' *Journal de la Société des Océanistes* 138/139: 37–51. doi.org/10.4000/jso.7118

CSIRO (Commonwealth Scientific and Industrial Research Organisation), 1996. *Review of Riverine Impacts: Porgera Joint Venture*. Canberra: CSIRO Environmental Projects Office.

Feld, S., 1996. 'Waterfalls of Song: An Acoustemology of Place Resounding in Bosavi, Papua New Guinea.' In S. Feld and K. Basso (eds), *Senses of Place*. Santa Fe (NM): School of American Research Press.

Filer, C. (ed), 1999. *Dilemmas of Development: The Social and Economic Impact of the Porgera Gold Mine, 1989–1994*. Canberra: Asia Pacific Press.

Foster, J.B., 2000. *Marx's Ecology: Materialism and Nature*. New York: Monthly Review Press.

Goldman, L., 1998. 'A Trickster for All Seasons: The Huli *Iba Tiri*.' In L.R. Goldman and C. Ballard (eds), *Fluid Ontologies: Myth, Ritual and Philosophy in the Highlands of Papua New Guinea*. Westport (CT): Bergin & Garvey.

Goldman, L.R. and C. Ballard (eds), 1998. *Fluid Ontologies: Myth, Ritual and Philosophy in the Highlands of Papua New Guinea*. Westport (CT): Bergin & Garvey.

Handley, G. and D. Henry, 1987. 'Porgera Environmental Plan: Report on Small Scale Alluvial Mining.' Porgera District Administration archives (unpublished manuscript).

Jacka, J.K., 2001a. 'On the Outside Looking In: Attitudes and Responses of Non-Landowners towards Mining at Porgera.' In B. Imbun and P. McGavin (eds), *Mining in Papua New Guinea: Analysis and Policy Implications*. Port Moresby: University of Papua New Guinea Press.

——, 2001b. 'Coca-Cola and *Kolo*: Land, Ancestors, and Development.' *Anthropology Today* 17(4): 3–8. doi.org/10.1111/1467-8322.00068

——, 2007. 'Whitemen, the Ipili and the City of Gold: A History of the Politics of Race and Development in Highlands New Guinea.' *Ethnohistory* 54: 445–471. doi.org/10.1215/00141801-2007-003

——, 2015. *Alchemy in the Rain Forest: Politics, Ecology, and Resilience in a New Guinea Mining Area*. Durham (NC): Duke University Press.

Johnson, P., 2011. 'Scoping Project: Social Impact of the Mining Project on Women in the Porgera Area.' Port Moresby: Porgera Environmental Advisory Komiti.

Kirsch, S., 2006. *Reverse Anthropology: Indigenous Analysis of Social and Environmental Relations in New Guinea*. Stanford (CA): Stanford University Press.

——, 2014. *Mining Capitalism: The Relationship between Corporations and their Critics*. Berkeley: University of California Press.

Kuletz, V., 1998. *The Tainted Desert: Environmental and Social Ruin in the American West*. New York: Routledge.

Marx, K., 1973. *Grundrisse: Foundations of the Critique of Political Economy* (transl. M. Nicolaus). New York: Random House.

——, 1990. *Capital: A Critique of Political Economy, Vol. 1* (transl. B. Fowkes). London: Penguin Classics.

May, R.J., and M. Spriggs (eds), 1990. *The Bougainville Crisis*. Bathurst (NSW): Crawford House Press.

Meggitt, M.J., 1964. 'Male–Female Relationships in the Highlands of Australian New Guinea.' *American Anthropologist* 66: 204–224.

Mittermeier, R.A., C.G. Mittermeier, T.M. Brooks, J.D. Pilgrim, W.R Konstant, G.A.B. da Fonseca and C. Kormos, 2003. 'Wilderness and Biodiversity Conservation.' *Proceedings of the National Academy of Sciences* 100: 10309–10313. doi.org/10.1073/pnas.1732458100

Robinson, N.C., 1960. 'Report of Extended Patrol in the Native Mining Area of the Porgera River Western Highlands, August–October 1960.' Unpublished manuscript.

Shearman, P., 1995. The Environment and Social Impact of the Porgera Mine on the Strickland River System. Hobart: University of Tasmania (Honours thesis).

——, 2001. 'Giving Away Another River: An Analysis of the Impacts of the Porgera Mine on the Strickland River System.' In B. Imbun and P. McGavin (eds), *Mining in Papua New Guinea: Analysis and Policy Implications*. Port Moresby: University of Papua New Guinea Press.

Talyaga, K., 1982. 'The Enga Yesterday and Today: A Personal Account.' In B. Carrad, D. Lea and K. Talyaga (eds), *Enga: Foundations for Development*. Armidale (NSW): University of New England, Department of Geography.

Vogt, C., 2012. *International Assessment of Marine and Riverine Disposal of Mine Tailings*. Nairobi: United Nations Environment Programme.

Wardlow, H., 2006. *Wayward Women: Sexuality and Agency in a New Guinea Society*. Berkeley: University of California Press.

Weiner, J.F., 1991. *The Empty Place: Poetry, Space, and Being among the Foi of Papua New Guinea*. Bloomington: Indiana University Press.

Wiessner, P. and A. Tumu, 1998. *Historical Vines: Enga Networks of Exchange, Ritual, and Warfare in Papua New Guinea*. Washington (DC): Smithsonian Institution Press.

Winchomba, K.A., 1970. 'Porgera Patrol Report 1 of 1969/70: Western Highlands District.' Port Moresby: Territory of Papua and New Guinea, Director of Native Affairs.

6. 'At Every Bend a Chief, At Every Bend a Chief, Waikato of One Hundred Chiefs': Mapping the Socio-Political Life of the Waikato River

MARAMA MURU-LANNING

Introduction

At 425 kilometres, the Waikato River is the longest river in New Zealand, and a vital resource for the country (McCan 1990: 33–5). Officially beginning at Nukuhau near Taupo township, the river is fed by Lake Taupo and a number of smaller rivers and streams throughout its course. Running swiftly in a northwesterly direction, the river passes through many urban, forested and rural areas. Over the past 90 years, the Waikato River has been adversely impacted by dams built for hydro-electricity generation, by runoff and fertilisers associated with farming and forestry, and by the waste waters of several major industries and urban centres. At Huntly, north of Taupiri (see Figure 6.1), the river's waters are further sullied when they are warmed during thermal electricity generation processes. For Māori, another major desecration of the Waikato River occurs when its waters are diverted and mixed with waters from other sources, so that they can be drunk by people living in Auckland.

Figure 6.1 A socio-political map of the Waikato River and catchment.
Source: Created by Peter Quin, University of Auckland.

As the Waikato River is an important natural resource, it has a long history of people making claims to it, including Treaty of Waitangi[1] claims by Māori for guardianship recognition and management and property rights.[2] This process of claiming has culminated in a number of tribes

---

1   The Treaty of Waitangi was signed by the British Crown and more than 500 Māori chiefs in 1840. It is the founding document of the nation state of New Zealand, laying out, in broad principles, the rights of both British and Māori. Many conflicts have arisen, however, over differing interpretations of these principles and over discrepancies between the English and Māori written versions of the document.
2   Over the last 35 years, New Zealand governments have provided opportunities for Māori tribes to seek redress for Crown breaches of the Treaty of Waitangi.

around the Waikato River signing deeds of settlement with the Crown between 2009 and 2012.[3] In return for their agreement not to pursue ownership claims to the river, the Crown has committed NZ$210 million to create a fund to assist in restoring the river's health and well-being. The major outcome of the combined agreements is a modern co-governance arrangement for the river that has equal Māori and Crown representation.

My contribution to this volume foregrounds the social life of the Waikato River (Figure 6.2), a river that is referred to as a *tupuna awa* (river ancestor) by Māori from the *iwi* (large tribes) of the Waikato region. In particular, I shall explore how Māori tribes and their leaders from the region have become prominent players in the river's politics and welfare. Because this work is primarily concerned with Māori engagement and relationships with the Waikato River, my focus is on the various *iwi* and *hapū* (sub-tribes) associated with the river, but I shall also discuss the roles that the Crown and the electricity company, Mercury (formerly Mighty River Power),[4] play in the river's socio-political landscape.

Figure 6.2 Children playing next to the Waikato River.
Source: Photo by Jeff Evans.

3   These have been legislated as the *Waikato-Tainui Raupatu Claims (Waikato River) Settlement Act 2010*; the *Ngāti Tūwharetoa, Raukawa, Te Arawa River Iwi Waikato River Act 2010*; and the *Ngā Wai o Maniapoto Ngāruawāhia Waipa River) Act 2012*.
4   The company's name was changed in 2016 in a move to emphasise its new focus on solar energy and electric vehicles.

I begin my study by mapping some of the historic relationships that *iwi* and *hapū* have with the Waikato River and with one another. Throughout its course, the Waikato River has many different characters and meanings for the people whose territories it flows through. There follows a conventional discussion of the formation of the Māori political movement called the Kīngitanga (the King Movement), which was founded on the banks of the Waikato River in the 1850s in response to British invasions of the Waikato region. I then touch on the confiscation of the lower Waikato River region by the British colonial government, and the impacts of this on Māori, before examining more recent controversies—most specifically the commodification of Waikato river waters through the development of hydro and thermal electricity generation. Finally, I address the recent partial privatisation of the electricity generating company, Mercury, and examine the contradictory implications of this initiative for Waikato-Tainui Māori. For example, some Māori understand that, when Mercury uses Waikato river waters to create power, what is in fact flowing through the company's turbines is the *tupuna awa* of Waikato *iwi*.

Māori Settlement of Lands Adjoining the Waikato River: From Taupo Nui a Tia to Te Puaha o Waikato

The Māori of the Waikato River belong to one of two tribal confederations: Te Arawa and Tainui. Figure 6.1 identifies the locations of the Te Arawa and Tainui tribes and communities. In this section, I build on previous ethnographic and historical works by Stafford (1967), King (1984), McCan (1990), Jones and Biggs (1995), Stokes (1997) and Grace (2005). Piecing together Māori claims to the river and occupation of the region, I shall share a number of oral traditions to demonstrate, first, the ways in which different *iwi* and *hapū* are connected to specific parts of the region and, second, to outline the ways in which *iwi* and *hapū* boundaries have been historically demarcated. A full account of Māori oral traditions of the Waikato River would include accounts of its origin, discovery and settlement; intertribal warfare and disputes over territory; tribal alliances through marriage and fighting; notable gatherings; prominent chiefs; the sharing and exchange of resources; and the curative power of the river's waters. While many of the oral traditions collected belong to specific tribal groups, some of the stories are shared by *iwi* and *hapū* that have similar interests in the river.

The Te Arawa Tribes

The central plateau containing the source of the Waikato River is a place with historical connection to the Te Arawa people. Te Arawa ancestors are understood to have come to New Zealand from a place called Hawaiki between 800 and 1,000 years ago, settling in the Rotorua and Taupo districts.[5]

Ngāti Tūwharetoa

Lake Taupo and the lands around it are recognised by Māori as the ancestral territory of Ngāti Tūwharetoa. The people of Ngāti Tūwharetoa descend from the powerful navigating chief Ngatoroirangi, who steered the Te Arawa canoe on its voyage from Hawaiki. Ngatoroirangi is credited with discovering and claiming the Lake Taupo region for his tribe.

In the 1920s, Lake Taupo and the Waikato River became essential resources in New Zealand's electricity development. At that time, the Crown refused to recognise Ngāti Tūwharetoa's ownership of the lake, and, in 1926, the New Zealand Government passed a law making Lake Taupo's bed the property of the Crown. To quell Ngāti Tūwharetoa's claim over the lake, the Crown agreed to an annuity for the tribe: a sum equivalent to 50 per cent of the gross revenue from the sale of the lake's trout fishing licences. However, Ngāti Tūwharetoa were further aggrieved when they were not consulted about the lake's outlet to the Waikato River at Nukuhau being altered for hydro-electric development in 1940, and also when waters from the Tongariro power scheme were diverted into Lake Taupo. Finally, in 1992, after years of disputing the Crown's method of acquiring the lake, an historic settlement agreement was reached when the Crown returned the ownership of the lake bed to Ngāti Tūwharetoa.

Ngāti Tūwharetoa is the fifth largest *iwi* in New Zealand; in the 2013 census, its population was recorded as 35,874. The tribe is reckoned to have at least 55 *hapū* and 81 *marae* (residential communities). There are at least 10 Ngāti Tūwharetoa *hapū* located around the Waikato River. After leaving Lake Taupo at Nukuhau, the river runs northeast to the Huka Falls or Hukanui (meaning 'great body of foam') and, after passing over the falls, it leaves the territory of Ngāti Tūwharetoa and enters that of Ngāti Tahu.

---

5   Hawaiki is the traditional Māori place of origin. It is where Io, the supreme being, created the world and the first people. It is also the place to which people return after death.

Ngāti Tahu-Ngāti Whaoa

With a territory that encompasses the 'great bend' of the Waikato River, Ngāti Tahu are recognised as the people of Lake Rotokawa (Stokes 2004: 53). Like Ngāti Tūwharetoa, the people of Ngāti Tahu are also descendants of the Te Arawa canoe (Grace 2005: 201). When the geographer Evelyn Stokes traced the history of Ngāti Tahu, she recorded that, in the nineteenth century, they were a nomadic people who had seasonal dwellings at Ohaaki, Orakei Korako and Nga Awa Purua (see Figure 6.1).

As the settlements were located in areas of geothermal activity (see Figure 6.3), it was unsafe for Ngāti Tahu people to live there permanently. The seasonal settlements called *nohoanga* were established for gathering prestige goods such as fish, birds, red ochre and sulphur (Stokes 2004: 55). In 1987, the Ngāti Tahu Tribal Trust wrote a submission to the Waikato Catchment Board, opposing a proposed sulphur mining operation in the area. The Trust wrote that prior to British settlement in the area:

> Their principal fern grounds were on Oruahineawe, on the north bank of the Waikato River, at Otamarauhuru and along the banks of the Parariki Stream. The Waikato River was a source of fish—kokopu (native trout) and inanga (whitebait), tuna (eels) and koura (fresh water crayfish). Kokopu were a particular speciality of the section of river below Nga Awa Purua. (Stokes 2000: 126)

Figure 6.3 Geothermal activity on the Waikato River.

Source: Reproduced from an 1885 Charles Blomfield oil painting entitled 'Orakei Korako on the Waikato', with permission from the Museum of New Zealand.

The submission explained that Nga Awa Purua was not only the name of an important seasonal settlement on the north bank of the Waikato River, but was also the name for that part of the river. Stokes (1987: 3) wrote that the name implies the 'head of navigation where the river divides and breaks into rapids'. As Ngāti Tahu only established temporary residences in this part of their territory, Stokes' research is important as it demonstrates Ngāti Tahu's longstanding connection to the Waikato River. In some official documents, Ngāti Tahu are identified as an *iwi* of Te Arawa. However, as a result of the Waitangi Treaty claims process and a desire to pursue their own claim separate from other Te Arawa *iwi* and *hapū*, Ngāti Tahu joined with Ngāti Whaoa, another *hapū* in the Reporoa area. The existing Te Puni Kokiri tribal membership figures record that these two groups have a combined membership of 2,724 people and are comprised of five *hapū* and six *marae*.

The Tainui Tribes

After flowing through Ngāti Tahu–Ngati Whaoa territory, the Waikato River arrives at Whakamaru, a place that is associated with the Tainui people of Ngāti Raukawa (Waitangi Tribunal 1993: 19–22). Like the Te Arawa canoe, the Tainui canoe or *waka* is said to have carried Māori ancestors from Hawaiki to Aotearoa (New Zealand). These people settled the northwestern quadrant of the North Island. The term Tainui *waka* also refers to the current confederation of Tainui *iwi*, which comprises Ngāti Maniapoto, Ngāti Raukawa, Ngāti Hauā, Hauraki, Ngāiterangi and the tribes of Waikato that were unified under the Kīngitanga. Oral tradition explains that Hoturoa was the captain of the Tainui canoe when it sailed from Hawaiki. When the canoe landed in Kawhia, its members settled in western parts of the central North Island. The boundary of Tainui territory is recited as:

Mokau ki runga	From Mokau in the south
Tamaki ki raro	To Tamaki in the north
Mangatoatoa ki waenganui	Mangatoatoa at the centre
Ki te kaokaoroa o Patetere	The long armpit of Patetere
Ki te Nehenehenui	The big forest of Maniapoto
Pare Waikato	From the mouth of the Waikato River in the west
Pare Hauraki	To all of Hauraki

Ngāti Raukawa

While Ngāti Tahu people contend that their ancestors, Tia and Ngatoroirangi, claimed the land on both sides of the river between Taupo and Atiamuri, the members of Ngāti Raukawa allege that they have special rights in the area because their ancestors, Wairangi and Whaita, who were great fighting chiefs, took possession of the territory from Ngāti Tahu by conquest. Today, the stretch of river between Atiamuri and Putaruru is mainly populated by these descendants of the Tainui ancestor Raukawa. The ancestral mountain of Ngāti Raukawa is Maungatautari, which is located near the Waikato River, just south of the township of Cambridge. There are debates over the tribal boundaries in Ngāti Raukawa territory. According to one informant who worked as a labourer on the Karapiro, Maraetai and Whakamaru dam projects:

> The Raukawa ancestors, Whaita and Wairangi, secured authority over land for Raukawa descendants along the Waikato River not just between Maungatautari and Whakamaru but all the way from Maungatautari to Lake Taupo. (Matangi Hepi, interview, May 1999)

The boundary recited by Matangi Hepi is also recorded as evidence in *The Pouakani Report* (Waitangi Tribunal 1993: 20), which provides accounts of the rival claims of the tribal groups that occupy the block of land known as Pouakani. This report echoes the account of Pei Te Hurinui Jones, a Tainui scholar with a strong interest in local history, who proposed that Ngāti Raukawa had tribal authority over the section of the Waikato River between Whakamaru and Maungatautari (ibid.: 21). From the interviews he conducted with Māori elders over many years, he inferred that, at some stage, Ngāti Raukawa moved up the Waikato Valley from Maungatautari, and then either displaced or absorbed a Te Arawa tribe, known as Ngāti Kahupungapunga, whose members were alleged to have lived between Putaruru and Atiamuri (Jones and Biggs 1995: 138).

With the establishment of pine forests around Taupo and Tokoroa in the 1920s, and the building of Kinleith Mill at Tokoroa in 1953, many Ngāti Raukawa members were employed in the forestry and pulp and paper industries. However, the down-scaling of Kinleith's timber processing and pulp and paper production in the 1980s resulted in much unemployment and financially difficult times for Ngāti Raukawa people. The current Te Puni Kokiri tribal membership figures record that there are 5,175 Ngāti Raukawa members living around the towns of Putaruru and Tokoroa.

Ngāti Koroki-Kahukura

While Mt Maungatautari is recognised as an ancestral mountain for Ngāti Raukawa, it is also the location of Ngāti Koroki-Kahukura, an influential *hapū* because of its enduring ties and shared boundaries with Ngāti Raukawa, Ngāti Maniapoto and Ngāti Hauā. The people of Ngāti Koroki-Kahukura are located along the middle reaches of the Waikato River. Genealogical accounts show that the ancestor Koroki was the father of the ancestor Hauā. As one of the members explained:

> Maungatautari acts as a *pou rahui* [boundary marker] for the tribe of Ngāti Raukawa and tribes that affiliate with Waikato-Tainui. It is an important gathering place for all the people. While everyone is welcome at Maungatautari, we affiliate to the Kīngitanga. (Karihana Wirihana,[6] interview, June 2001)

While much of Ngāti Koroki-Kahuhura's ancestral lands were not included in the Crown's mass land confiscations of 1863, the tribe is notable because it took in Waikato people who were exiled from their own lands after the confiscations. While Ngāti Koroki-Kahukura was recognised as an *iwi* in its own right prior to the confiscation of the southern part of Waikato district, in Treaty of Waitangi claim contexts it is generally regarded as being a part of Waikato *iwi*. Nevertheless, Ngāti Koroki-Kahukura filed a Waitangi Tribunal claim to protect their interests in the Waikato River, and members contend that they have authority for the river between Arapuni and Karapiro. According to their treaty negotiator, Willie Te Aho, the area of the river that Ngāti Koroki-Kahukura are claiming is outside Waikato-Tainui's tribal territory (Anon. 2008).

Ngāti Hauā

The section of the Waikato River from Mt Maungatautari to Horotiu, or from Pukerimu to Ngāruawāhia, is understood to be part of Ngāti Hauā's ancestral territory (Jones and Biggs 1995: 334). Until the mid-nineteenth century, Ngāti Hauā was recognised as a formidable *iwi* by its Māori counterparts, and its lands were envied and revered for their rich soils. When the British invaded the Waikato region in the 1860s, the section of the Waikato River occupied by this group was called the Horotiu River (Belich 1998: 164, 179; Stokes 2002: 14). Much of their

---

6   Karihana Wirihana lives in Cambridge and is from Maungatautari Marae. He was the captain of the Waka Rangatahi, which is part of the fleet of Tainui canoes associated with the late Princess Te Puea Herangi.

territory was then confiscated, including lands in the Matamata, Hamilton and Morrinsville areas. Like Ngāti Koroki-Kahukura, the people of Ngāti Hauā are generally regarded as part of the Waikato *iwi*.

Ngāti Mahuta

Ngāruawāhia is one of many places customarily associated with Ngāti Mahuta, and is the location of the large Waikato *marae* known as Tūrangawaewae. The place known as 'The Point' in Ngāruawāhia is the place where the Waikato River joins the Waipa River (see Figure 6.4).

Figure 6.4 The place or 'point' where the Waipa River and the Waikato River meet.
Source: Photo by Jeff Evans.

Oral traditions explain that the merging of the two rivers in Ngāruawāhia symbolises the union of Ngāti Raukawa, Ngāti Maniapoto and Waikato people through the marriage of Ngawaero, who was of Ngāti Raukawa-Maniapoto descent, and Te Wherowhero, an important Ngāti Mahuta chief. As a result of the marriage, the Waipa River is sometimes personified as Ngawaero, and the Waikato River is personified as Te Wherowhero. Jones described the relationship as follows:

> There are certain places in Aotearoa that seem to have a spell of strength and endurance cast upon them by primeval forces. In the Waikato is one of these places where the waters of the Waikato and Waipa Rivers meet at Ngāruawāhia. For more than 12 miles above this point the clear and deep flowing waters of the Waikato, in its westward course, appear to be bent on charging straight through the outflung Hakarimata range. But as if in obedience to the quiet persuasion of the sluggish waters of the Waipa, it turns with renewed zest to the north, where for about three miles it rushes through Taupiri gorge to the west of Taupiri Mountain. (Te Hurinui 1959: 134)

Approximately 7 km north of Ngāruawāhia is Mt Taupiri, an important burial place and ancestral mountain for Waikato Māori, who believe the Waikato River to be very sacred, as it flows past this landmark. There is a story shared by Waikato and Ngāti Tūwharetoa Māori regarding the origin of the river. It was told by Ngāti Tūwharetoa's paramount chief, Sir Tumu Te Heuheu, to guests of the late paramount chief of Waikato, Te Arikinui Te Atairangikaahu,[7] who were attending the 40th anniversary of her coronation at Tūrangawaewae Marae in 2006.[8]

Tongariro and Taupiri were a brother and a sister mountain that lived in the Central Plateau. The siblings were very close. At birth Taupiri was promised to the Tainui mountain chief, Pirongia. On reaching adulthood, Taupiri married Pirongia and went to live in his territory. At her new home, Taupiri became very sick. Concerned by his wife's condition, Pirongia gathered the most knowledgeable healers to treat Taupiri. However, none of the healers were able to cure her. Aware that her people had a remedy, Taupiri asked Pirongia if he would send word of her illness to Tongariro. Desperate for his wife's health to improve, Pirongia asked a trusted servant to make the difficult journey to Taupiri's homeland. The servant agreed and took his dog as a companion.

After several days the servant and the dog arrived at Tongariro's village. On their arrival they were welcomed and fed, and then taken to meet Tongariro. Tongariro explained that they needed to get a good night's sleep since, before dawn the next morning, they would be climbing a great mountain to obtain a cure for Taupiri. The next day Tongariro, the servant and his dog made their ascent. Stopping at a special place on the mountain, Tongariro recited a chant and then struck a rock with his walking stick. Suddenly, pure water emerged from the rock. Using calabashes that they had brought with them, the servant collected some of the special water for Taupiri. After they had filled the calabashes with the water, the servant thanked Tongariro, and then he and the dog began their long journey home. As the pair departed, Tongariro instructed the water, which had become a stream, to follow them so that Taupiri would always have a supply of the sacred water at her disposal.

On their way home the servant and the dog passed through a gigantic crater that the stream filled. This crater we now know as Lake Taupo-nui-a Tia. At Tapuaeharuru (near Nukuhau), a place at the northern end of the lake, the stream turned into a powerful river. Some versions of the story say that the Te Arawa people tried to entice the river to flow

---

7   The title used by Waikato Māori when referring to the paramount chief Te Atairangikaahu.
8   The following text is a version of the speech that I recorded in writing as it was being made.

through their lands, but at Te Ohaaki the dog dug a ditch, preventing the river from going in their direction. However, at Piarere, a place between Tirau and Karapiro, the river became distracted when it heard the call of its sea-parents. Unable to resist their voices, the river turned down into the Hinuera Valley and made its way eastward across the Hauraki plains where it met up with its parents at the Thames estuary. As the servant and the dog were unable to persuade the river to follow them, they continued home with the calabashes of sacred water.

On their return Taupiri drank the water and recovered immediately. The servant then told Taupiri of Tongariro's gift—the stream that had become a runaway river. On hearing the story, Taupiri began to chant. When Tongariro heard his sister's call he also began to chant. In unison their calls woke Ruaumoko, the deity of earthquakes. Ruaumoko was angry at being woken, and his fury caused the earth to shake and split and volcanoes to erupt. The runaway river, knowing that it was the cause of Ruaumoko's anger, immediately diverted its course so that it could be at Taupiri's disposal. When the river reached Taupiri she explained that it was now free to go and be with its parents. Flowing westward the river was reunited with its sea-parents at Port Waikato.

This story emphasises the unique bond between Waikato and Ngāti Tūwharetoa people. Tongariro and Taupiri not only represent two gendered chiefly ancestors, but also the peoples of Ngāti Tūwharetoa and Waikato. The Waikato River, a gift from Ngāti Tūwharetoa to Waikato people, is portrayed as a young, non-gendered waterway with curative powers. However, Waikato *iwi* also refer to the Waikato River as their primordial *tupuna awa*.

Beyond Taupiri, the Waikato River flows past several *marae* settlements. Waahi Marae in Huntly is an important one because it has been the home of many Kīngitanga leaders, and there are others between Huntly and the mouth of the river at Port Waikato (or Te Puaha o Waikato) that are significant to Waikato Māori identity (see Figure 6.1). Before the Waikato land confiscations in the 1860s, all land in the region belonged to autonomous *iwi* and *hapū* groups, but, once the lands were confiscated, they were subdivided and then turned into farms by British colonists. The *marae* dotted along the river are the only visible symbols of prior Māori occupation of the region. The 2013 census recorded the combined population of all the Waikato-Tainui *iwi* as 40,083 people, divided between 33 *hapū* and 160 *marae*.

The Kīngitanga Emerges

Māori social organisation around the Waikato River was transformed with the arrival of British colonists. The first Europeans began visiting the Waikato Valley in the 1820s, though it was another 20 years before they settled in the area. Throughout the 1840s, Māori agriculture flourished in the Waikato region as *hapū* groups cultivated a variety of crops for the growing town of Auckland. The Waikato River was an important element of their success because it provided a reliable transport route to markets in the north, and their success did not go unnoticed. Stokes (1997: 10) has explained that 'the productivity of the Waikato lands, especially in the Hamilton basin and Waipa Valley, attracted the attention of British settlers, officials and land speculators'. The identity of Waikato-Tainui is closely linked with the river and the region. Many *whakataukī* ('proverbs') refer to the river or the surrounding region. Tūkino Te Heuheu I, a paramount chief of Ngāti Tūwharetoa, expressed this proverb to acknowledge Pōtatau Te Wherowhero during the search for a Māori king in the 1850s:

Ko Waikato te awa	Waikato is the river
Ko Taupiri te maunga	Taupiri is the mountain
Ko Te Wherowhero te tangata	Te Wherowhero is the chief
Waikato Taniwharau	Waikato of a hundred chiefs
He piko, he taniwha	At every bend, a chief
He piko, he taniwha	At every bend, a chief

The Kīngitanga movement was established in 1858, led by Pōtatau Te Wherowhero of Ngāti Mahuta, to resist the appropriation of Māori land. Pōtatau was chosen to be the Māori king 'because of his reputation as a fighting chief and high genealogical status which linked him to Hoturoa the Captain of the Tainui canoe' (Mahuta 1975: 1). In addition, Kamira Binga Haggie of Tūrangawaewae Marae has pointed out that Pōtatau was selected because of his amicable relationships with the many tribes of the lower Waikato River, who could supply food for massive Kīngitanga gatherings (Haggie 1997).[9] Pōtatau only survived for two years as Māori monarch; when he died, supposedly at the age of 85, he was succeeded by his oldest son Matutaera Tāwhiao Te Wherowhero—King Tāwhiao.

---

9   Kamira Binga Haggie was the chairperson of the Tūrangawaewae Marae Committee for many years and was a tribal spokesperson for the Waikato River. He was a keen rower, whitebaiter and duck shooter.

Māori refusals to sell did not curb the Crown's desire to own large areas of fertile land, so the Crown's representative, Governor Grey, sought new methods to obtain land for British colonists in the Bay of Plenty, Taranaki and Waikato regions. The invasion of the Waikato region commenced on 11 July 1863 (Belich 1998: 119). It involved a series of assaults as far south as Ngāruawāhia. The most debilitating of these took place at Meremere and Rangiriri (see Figure 6.5), where the British militia used flotillas of vessels to attack Māori settlements (ibid.: 145). These attacks shattered the livelihoods of Waikato Māori as people were killed, lands were seized, and *marae* and tribal canoes were destroyed. The capture of Rangiriri provided a straightforward entry into the rest of the Waikato region, enabling General Cameron, who was leading the invasion, to take King Tāwhiao's headquarters at Ngāruawāhia.

Figure 6.5 Sign at Rangiriri marking the British militia's invasion of Waikato *iwi* territory in 1860.
Source: Photo by Jeff Evans.

In the 1860s, the colonial government created two pieces of legislation to remove *hapū* from their lands. First, the *Suppression of Rebellion Act 1863* proclaimed that Waikato and a number of other tribes were engaged in a rebellion against the Crown and that, as punishment for these actions, their lands would be confiscated. The confiscated lands were then given to British settlers in the form of grants made under the *New Zealand Settlements Act 1863*. While some of the confiscated lands were given as payments to men who had fought for the Crown against the Waikato tribes, a substantial amount of land was also sold to incoming British settlers. Small blocks of poor-quality land were set aside for returning rebel Māori so that they could establish reserves once they gave their allegiance to Queen Victoria, but few Waikato Māori were interested in these unproductive lands (McCan 2001: 58).

The British view was that, when Māori lands adjoining the river were confiscated, their rights in the Waikato River were also removed. The signing of the Treaty of Waitangi in 1840 had brought Māori under the jurisdiction of the Crown, so that they were governed by English common law. Under English common law, the owner of land adjacent to a non-tidal river possessed a right to use the water, and the landowner also held a property right over the portion of the riverbed extending to the mid-stream point in the river. Therefore, when Māori land adjacent to the river was confiscated, Māori also lost their common law rights to the riverbed.[10]

The alienation of Māori from the Waikato River worsened when the Crown appropriated the rights to coal deposits beneath the settlement of Huntly. To establish coal production in the area, the Crown needed to own land in the Huntly area and exercise control over the Waikato River because the waterway was necessary to transport the coal to Auckland. The Crown therefore extinguished all Māori and non-Māori property rights in the Waikato River by enacting the *Coal Mines Amendment Act 1903* for what was said to be the public good. Stokes (2004: 48) has explained that, under this Act, the *ad medium filum aquae* rule was replaced by a declaration of Crown ownership of all the beds of 'navigable' rivers in New Zealand.

The confiscation of Waikato lands affected Māori from the region containing the middle and lower reaches of the Waikato River, and the Ngāti Hauā people around Horotiu and in the Waharoa area around Matamata (see Figure 6.1). A number of the tribes newly affiliated to the Kīngitanga were later forced to leave their lands and retreat to those of their closest allies.

---

10   This English law known as the rule of *ad medium filum aquae* was first recognised in New Zealand in the case of *R v Joyce* (1906) 25 NZLR 75.

Princess Te Puea Herangi Establishes Tūrangawaewae Marae

The third Māori king, Mahuta, relied heavily on the foresight and hard work of his niece, Te Puea Herangi (King 1984: 20). She was a granddaughter of King Tāwhiao through her mother, Tiahuia. Te Puea first gained prominence as a leader when she led a campaign against the conscription of Waikato Māori during World War I. She argued that there was no point in fighting for a country in which her tribe had no land. Te Puea affirmed her standing as a leading figure of the Kīngitanga in 1921, when she left her home in Mercer on a barge with a party of workers to rebuild Tūrangawaewae Pa at Ngāruawāhia, which is one of the places where King Tāwhiao lived prior to being exiled from Ngāti Mahuta lands.

The revival of Waikato *iwi* and the Kīngitanga thus began with the construction of Tūrangawaewae Pa on the banks of the Waikato River in the 1920s (see Figure 6.6). Prior to the confiscation of Waikato lands, the original *pa* (traditional village) in Ngāruawāhia was known as Pikiarero and was located at 'The Point', which is the junction of the Waikato and Waipa rivers (see Figure 6.4). However, Te Puea was unable to re-acquire that land, so land was bought on the other side of the Waikato River. Not only did Te Puea establish Tūrangawaewae Marae, which was a great feat in itself, but she also set up a carving school within the new settlement, built a series of meeting houses and other community facilities throughout the Waikato region and King Country,[11] composed songs, trained the Māori concert party known as Te Pou o Mangatawhiri and established the Tainui Māori Trust Board. The latter was established in 1947 to administer money that Te Puea negotiated from the government as a form of compensation for confiscated Waikato lands (King 1984: 338). Te Puea's influence in the Kīngitanga lasted almost 40 years, during which time she made explicit the relationship that Waikato Māori and the Kīngitanga had with the Waikato River, as exemplified in 1929 when she rallied support for a petition for the return of tribal fishing rights in the river (Orange 2004: 122).

---

11   The term 'King Country' resulted from the invasion of Waikato by the colonial forces in the 1860s. After the attack, Kīngitanga members were forced south of what was called the *aukati*, or boundary, seeking refuge in a line of *pa* alongside the Puniu River near Kihikihi. Land behind the *aukati* remained Māori territory; Europeans were warned that if they crossed the *aukati* their safety would not be guaranteed.

Figure 6.6 Kapa Haka on the Waikato River at Tūrangawaewae Marae.
Source: Photo by Jeff Evans.

Te Puea specifically encouraged the people of Tūrangawaewae Marae to bring into play a number of cultural features that demonstrated their connection with the river. King recorded that she established the *marae* on the banks of the river:

> so that its waters would be a constant and reassuring presence. By the mid-thirties she was conscious that there was only one major way in which she was not making use of the river—Waikato no longer had any large canoes. (King 1984: 206)

Evidently, nothing had stirred Te Puea more in her youth than the sight of paddlers in Mahuta's ornately decorated canoe, Tahereheretikitiki. In the mid-1930s, she initiated the building of a fleet of ceremonial canoes to honour the Waikato River (see Figure 6.7). However, as Te Puea was creating Tūrangawaewae Marae, major developments were taking place in the electricity-generating industry along the middle and upper reaches of the river.

Figure 6.7 Rangatahi Waka on the Waikato River, captained by Karihana Wirihana.

Source: Photo by Jeff Evans.

Electricity Generation

The electricity industry was first established on the Waikato River at Horahora in 1913, when a dam was constructed by the Waihi Goldmining Company to supply power for the Waikino battery near Waihi (Stokes 1997: 46). This generating station was acquired by the government in 1919. The first state-initiated power development on the river occurred in 1929, when a dam and power station were built at Arapuni, in Ngāti Raukawa territory. Throughout the 1930s, geologists surveyed the gorges of the Waikato River and identified 10 more potential dam sites. The major control gates that now regulate the flow of the Waikato River were installed at Nukuhau on Lake Taupo between 1940 and 1941. These gates control the amount of water flowing down the Waikato River to meet electricity generation and regulatory requirements. The land required for dams, power stations, switchyards and other works, as well as land flooded by dam reservoirs, was taken under the Public Works Act in a series of proclamations between 1949 and 1982. The electricity developments on the Waikato River resulted in the construction of eight dams and reservoirs and nine hydro power stations.[12] Electricity developments that occurred in Ngāti Tūwharetoa, Ngāti Tahu, Ngāti Raukawa, Ngāti Koroki-Kahukura and Waikato territories resulted in the permanent loss of Māori lands and river access. Construction of dams and generating stations also caused serious disruptions to Māori settlement patterns in other areas. Matangi Hepi from Ngāti Raukawa, who worked on a number of these projects, recalled that:

> There was lots of labouring work on dam construction sites for local men, it was known that once a dam was complete the men were no longer necessary to NZED [New Zealand Electricity Department] and we left to find work elsewhere. Back then we could just move on to the next dam being built, but after Maraetai they stopped building them, a lot of local men and their families had to leave the area. (Matangi Hepi, interview, May 1999)

Dams and reservoirs now divide the Waikato River into sections, and have altered its aquatic ecosystems. Dams prevent eels, which have always been an important food source for Māori, from being able to swim downstream to spawn in the open sea. In the reservoirs behind the dams, fish such as the *kokopu* (native trout), which were plentiful in the upper reaches of the Waikato River, have been replaced by exotic species that are better adapted to lakes.

The construction of dams and reservoirs has resulted in parts of the Waikato River becoming permanently inaccessible. For instance, it is no longer possible to traverse the Aratiatia rapids because the Aratiatia dam gates are opened every two hours between 10 am and 2 pm or 4 pm each day,[13] allowing built-up water to surge down the narrow gorge for 30 minutes before being closed again. At the time of its construction there was no survey of archaeological sites, and many places along the river banks that were special to Māori were flooded. According to *The Pouakani Report*, these losses included:

---

12   The hydro-power stations known as Horahora #1 and #2 (1913) were followed by those at Arapuni (1929), Karapiro (1947), Whakamaru (1949), Atiamuri (1953), Ohakuri (1955), Waipapa (1961), Aratiatia (1964), Maraetai (1970), and two thermal power stations, Meremere (1950) and Huntly (1985). The Horahora stations were submerged when the larger Karapiro station was established (see Figure 6.1).

13   The last daily flow is at 2 pm in summer and 4 pm in winter.

the loss of the 'Waipapa rock paintings' at the confluence of the Waipapa and Waikato rivers. Other rock paintings submerged by Lake Arapuni. Hot springs at various places, such as Waimahana have been submerged. Two thirds of the active geothermal areas including geysers and papakainga at Orakei Korako, and hot springs and wahi tapu at Te Ohaaki, were submerged by the Ohakuri hydro lake. (Waitangi Tribunal 1993: 294)

Joseph Haumaha from Ngāti Raukawa lives next to Maungakaretu Marae at his family homestead in Putaruru West, which is close to the Arapuni power station. Electricity development in this area has not only restricted access to parts of the river, but has also changed the flow and level of the river's water. Although it has reduced the number of eels and fish in the river, Joseph explained that his family still regularly fish for eels and brown trout while also gathering *koura* (freshwater crayfish) from the river's tributaries.

> The boys and the nephews get brown trout at Arapuni. You can't get rainbows because the water is too muddy, they like to be nearer the lake. The dams have made a big difference to our fishing, there's no native fish now and the eels aren't fat anymore. You are supposed to have a licence to fish for trout around here — that's another thing we're not happy about. (Joseph Haumaha, interview, January 2004)

While Māori at the northern end of the Waikato River do not have to contend with some of the impacts of dam construction, Fookes (1976) reported the frustration and distress that Waahi Marae Māori experienced when the Huntly power station was built in 1973, including many of the impacts commonly associated with involuntary resettlement of Indigenous peoples.

Privatising Mighty River Power

In 1986, the New Zealand Government announced that it intended to reform both the generation and transmission sectors of the electricity industry. The first of the reforms occurred in 1987, when the Electricity Corporation of New Zealand (ECNZ) was set up as a state-owned enterprise (SOE) to own and operate the then Ministry of Energy's generation and transmission assets. Subsequently, ECNZ created a subsidiary called Transpower to run its transmission assets. The reforms continued in 1993 under the *Electricity Act 1992*, which effectively removed the statutory monopolies of existing distributors and the obligation to supply electricity, while the Electricity Market Company Ltd was established to facilitate a market framework for wholesale trading. In 1994, Transpower was separated from ECNZ and set up as a stand-alone SOE. The year 1995 saw ECNZ split into two competing SOEs: ECNZ and Contact Energy. In 1998, a range of reforms was announced by the government, including the sale of Contact Energy, and a further division of ECNZ into three rival SOEs. Of the three companies formed, two were dependent on the Waikato River for their generation processes. These were Genesis Energy, which operates the power station in Huntly, and Mighty River Power (MRP), a hydro generator with eight power stations along the Waikato River.

Figure 6.8 Waka Ama, a racing canoe co-sponsored by Mighty River Power and Waikato-Tainui.
Source: Photo by Jeff Evans.

In 2011, the national government proposed the partial privatisation of New Zealand's remaining state-owned electricity generating assets. When this was announced, some Māori leaders were keen to buy shares (Tahana 2011). They included Waikato leader Tukoroirangi Morgan, who announced that his *iwi*, Waikato-Tainui,[14] was excited by the prospect of buying shares in MRP (see Figure 6.8), which produces around

---

14   Waikato *iwi* came to be known as Waikato-Tainui after its 1995 land treaty settlement with the Crown.

16 per cent of New Zealand's electricity. To reassure New Zealanders that the assets would be in safe hands, Morgan explained that 'iwi won't sell [their shares], the investments are intergenerational' (Gifford 2011). However, Morgan and the leaders of other *iwi*, such as Ngāpuhi and Ngāi Tahu, changed their positions when the New Zealand Māori Council challenged the government's privatisation strategy and water ownership policy.[15] The council members, who regard themselves as guardians of the resources in which Māori have interests, warned that Māori should reject the asset sales until their property rights in fresh water had been determined (Māori Council 2013).

In September 2012, many Waikato River tribes were separately contacted by Crown representatives to discuss how the partial privatisation of MRP might affect them. These Crown–*iwi* meetings occurred at the same time that Waikato-Tainui's paramount chief King Tuheitia and his representative, Tukoroirangi Morgan, called a pan-tribal forum at Tūrangawaewae Marae to discuss Māori property rights in fresh water. While Māori leaders from all over the country attended the water forum, Prime Minister Key banned Crown representatives and National Party MPs (including Māori MPs) from attending the gathering. The concern for Māori was that the privatisation of MRP could hinder their future claims to property rights in fresh water. However, against the wishes of many Māori and non-Māori New Zealanders, MRP was partially privatised in May 2013.

It has been argued that the process of privatisation serves 'first to undo the relations between things and people, and then to remake them in a different form' (Alexander 2004: 48). Following this logic, the partial privatisation of MRP could transform the Māori relationship to the Waikato River by turning the water into a commodity divorced from its relationship with the land over which it flows, as well as from the Māori people themselves. Arguably, when the government privatised the company's power stations, turbines and dams, they privatised the waters of the Waikato River as well. In an earlier work, I argued that the New Zealand Government's current focus is on the economically important parts of the Waikato River—that is to say, on the water rather than on the banks or the riverbed (Muru-Lanning 2010).

---

15   The New Zealand Māori Council is a statutory body with elected representatives of all Māori people. The functions of the Council are set out in the *Māori Community Development Act 1962*.

The process of commodification has been examined by a number of scholars. One who reported on the commodification of water in the River Stour in England wrote that 'each new process, each layer of complexity, each investment of labour and knowledge, has transformed (water) from a "raw" or "natural" substance into a product' (Strang 2004: 36). Appadurai's edited contribution, *The Social Life of Things*, is another important work on the theme of commodification (Appadurai 1986: 6). In that volume, Kopytoff, in his discussion of slaves as a commodity, argues that the process of commodification involves a succession of phases during which the slave is sometimes a commodity and sometimes not. In Kopytoff's words:

> Slavery begins with capture or sale, when the individual is stripped of his previous social identity and becomes a non-person, indeed an object and an actual or potential commodity. But the process continues. The slave is acquired by a person or group and is reinserted into the host group, within which he is resocialized and rehumanized by being given a new social identity. (Kopytoff 1986: 65)

In other words, slaves are only commodities between the time they are captured and sold. Kopytoff's discussion of 'slaves as commodities' has relevance for analysing the processes through which the Waikato River has been turned into a commodity. Strang went on to argue that 'just as pricing reduces and commodifies "nature", the material culture of metering imposes human agency onto water emphasising its re-creation as a product, manufactured by the water industry' (Strang 2004: 228). For water to be commodified, it is necessary to physically and conceptually separate it from its riverine and cultural context, and then create instruments for measuring and pricing it. These processes of confining water and treating it as a commodity are not unique to water utilities, because controlling water is also crucial to the production of electricity, as in the case of the Waikato River.

MRP, now known as Mercury, captures water using control gates at Nukuhau and a series of dams along the river. River waters held in reservoirs behind the dams become lakes. These volumes of water are contained at different points on the river. Water held in lake storage systems is monitored and controlled by dispatchers who work in a trading room at the company's generation office in Hamilton (Titchall 2008: 12–13). River water is only a commodity from the time it is held in lake storage to the time it flows through the company's electricity-generating turbines. Indeed, it is not possible to commodify river water without the dam structures that capture the water. When water is turned into a commodity, it is easier to detach it from a past history that includes its relationships with different groups of people. Commodification does not fit with a Māori worldview because of the way that it conceptually separates things. In contrast, a Māori worldview is always seeking to connect one thing to another, and especially to connect people, lands and resources to Papatuanuku (Mother Earth), which is understood to be a living entity (Walker 2004: 11–14; Smith 1999: 74).

Voices from the Waikato River

In 2011 and 2012, I asked Waikato-Tainui members how they felt about the privatisation of MRP and their opportunity to become shareholders in the company. Their responses fell into four categories. First, there were those people who said they did not know the company was being sold. People from this group explained that the sale would not make any difference to them because they had no say in Waikato-Tainui or New Zealand politics. Second, there were humorous or sarcastic responses, like that of a 34-year-old male who had lived most of his life at Tūrangawaewae Marae: 'One minute it's a *tupuna* (ancestor), then it's an *awa* (river), next thing you know it's *apis* (ready cash) in the pocket'.[16] Another example would be this rhetorical question from a male in his 50s: 'Why bother when Ngāi Tahu's going to buy it?'[17] Third, were many Waikato-Tainui members who responded to my question with silence, which I would interpret as the most culturally appropriate response from those deeply affiliated with the Kīngitanga. Finally, there were some members who, like Tukoroirangi Morgan, believed that the tribe should buy shares. One female member, who has sat on Waikato-Tainui's tribal authority for some years, expressed her opinion as follows: 'Why not buy shares in Mighty River Power? It is better that *tangata whenua* [local Māori] buy them rather than the Americans or Chinese'.

---

16 The term *apis* is a neologism that derives from the name of Sir Apirana Ngata, a Māori leader whose face appears on the New Zealand 50-dollar note.
17 Ngai Tahu is an *iwi* with ancestral territory in the South Island, and is recognised as an influential tribe because of its members' financial success since signing their 1997 treaty settlement.

A similar response came from a male elder:

> The government's going to sell Mighty River Power whether we like it or not, that's the sort of government we have. If we don't buy shares, these companies will be sold to overseas buyers and then where will we be? We have an obligation to protect our interests in the *awa*—if we don't buy we'll get left out.

However, within a year of my interviews, the tribe's newly elected executives and their business advisors, who did not include Tukoroirangi Morgan, decided that Waikato-Tainui *iwi* would in fact *not* buy shares in MRP when the company was floated in 2013. Mike Pohio, the chief executive of Tainui Group Holdings, declared in a media statement that the tribe had 'no interest in asset share sales'. And, for many Waikato Māori, the final decision not to buy shares in the company came as no surprise. As Waikato lawyer, Shane Solomon, put it in October 2012:

> Why the hang would we want to buy shares in something we already own? We have a claim on the water. Mighty River should be paying us to use our water.

Conclusion

Overall, this study is characteristic of the way that Māori tribes are represented in Waitangi Tribunal reports and published tribal histories (Stafford 1967; Jones and Biggs 1995; Grace 2005). One purpose has been to describe the Waikato River as a setting, using a traditional Māori trope for representing relationships between people and territory. In constructing a commentary that highlights the relationships of the river to Māori as a cultural resource and boundary-making entity, my aim has been to show that the river's contested nature comes from the diverse range of 'interests' that people have in it. The Waikato River has a rich history of being claimed and contested because it is a very important resource. Yet, what I have also tried to show is that the river has multiple social lives that operate in unison. For now, the co-governance agreement with the Crown is the primary mechanism used by Māori to stake their interests and claims in the river. However, while the co-governance arrangement forces the local tribes and the Crown to work together for the well-being and betterment of the river, it should not be forgotten that each *iwi* and *hapū* has its own set of goals and aspirations for the Waikato River. Furthermore, Māori with interests in the river have found that the state's shedding of responsibility in certain areas of governance has not been accompanied by any relinquishment of real control over the important water resource.

While it is evident from the media coverage that leaders of the Waikato-Tainui and the Kīngitanga were annoyed by the government's process and rhetoric, I suggest that a more important ideological factor influenced Waikato-Tainui's decision not to buy shares in MRP. Since all but one of the company's power stations on the Waikato River are outside the tribe's ancestral territory (the exception being Karapiro), the *iwi*'s decision-makers may have felt that shares in the company were not vital to the tribe's identity and future aspirations. A second and more pragmatic reason may have been that, after conducting business evaluations, the tribal decision-makers felt that buying shares in the company was simply not a good investment, and could become problematic in the future if their own tribe, or a Māori collective of tribes, were to pursue ownership claims to fresh water.

However, what is also evident from expressions of disinterest in the asset sale, or from their joking, sarcasm and silence, is that a number of Waikato-Tainui Māori were not prepared to sell their *tupuna awa*. In this respect, and most importantly, this study develops ideas that may serve to 'root' Māori groups (*iwi, hapū, whānau* and *marae*) to particular lands and resources. It emphasises significant tribal histories and provides insight into the social life of the Waikato River as a *tupuna*.

References

Alexander, C., 2004. 'Value Relations and Changing Bodies: Privatisation and Property Rights in Kazakhstan.' In K. Verdery and C. Humphrey (eds), *Property in Question: Value Transformation in the Global Economy.* Oxford: Berg.

Anon., 2008. 'Iwi Awaits River Word.' *Waikato Times*, 4 September.

Appadurai, A. (ed.), 1986. *The Social Life of Things: Commodities in Social Perspective.* Cambridge: Cambridge University Press. doi.org/10.1017/CBO 9780511819582

Belich, J., 1998. *The New Zealand Wars.* Auckland: Auckland University Press.

Fookes, T., 1976. 'Social and Economic Impacts Feared by the Maori Communities at Huntly, 1972–73.' Hamilton: University of Waikato, School of Social Sciences, Huntly Social and Economic Impact Monitoring Project (Research Memorandum 4).

Gifford, A., 2011. 'Iwi Staking Their Claim.' *New Zealand Herald*, 9 August.

Grace, J., 2005 [1959]. *Tuwharetoa*. Wellington: A.H. and A.W. Reed.

Haggie, K.B., 1997. 'Tūrangawaewae Marae.' Video-recorded interview with Kamira Binga Haggie shown at Te Papa National Museum, Wellington.

Jones, P. and B. Biggs, 1995. *Nga Iwi o Tainui: The Traditional History of the Tainui People*. Auckland: Auckland University Press.

King, M., 1984 [1977]. *Te Puea: A Life*. Auckland: Hodder and Stoughton.

Kopytoff, I., 1986. 'The Cultural Biography of Things: Commoditisation as Process.' In A. Appadurai (ed.), *The Social Life of Things: Commodities in Cultural Perspective*. Cambridge: Cambridge University Press. doi.org/10.1017/CBO9780511819582.004

Mahuta, R., 1975. 'Taupiri Mountain: Submission to Support Application for Vesting Order in the Matter of the Land known as Allotment 458, 481 and 778 Parish of Taupiri.'

Māori Council, 2013. 'Water Claim Statement by NZMC & the Claimant Management Group.' 28 February. Viewed 6 June 2018 at: maoricouncil.com/2013/03/01/water-claim-statement-by-nzmc-the-claimant-management-group/

McCan, D., 1990. 'Water: Towards a Bicultural Perspective.' Hamilton: University of Waikato, Centre for Maori Studies (Information Paper 23).

——, 2001. *Whatiwhatihoe: The Waikato Raupatu Claim*. Wellington: Huia Publishers.

Muru-Lanning, M., 2010. Tupuna Awa and Te Awa Tupuna: An Anthropological Study of Competing Discourses and Claims of Ownership to the Waikato River. Auckland: University of Auckland (PhD thesis).

Orange, C., 2004. *An Illustrated History of the Treaty of Waitangi*. Wellington: Bridget Williams Books. doi.org/10.7810/9781877242168

Smith, L.T., 1999. *Decolonising Methodologies: Research and Indigenous Peoples*. London: Zed Books.

Stafford, D., 1967. *Te Arawa: A History of the Arawa People.* Wellington: A.H. and A.W. Reed.

Stokes, E., 1987. 'Public Policy and Geothermal Energy Development: The Competitive Process on Maori Lands.' In *Proceedings of the Symposium on New Zealand and the Pacific: Structural Change and Societal Response.* Hamilton: University of Waikato.

———, 1997. 'Ko Waikato te Awa: The Waikato River.' In E. Stokes and M. Begg (eds), *Te Hononga ki te Whenua, Belonging to the Land: People and Places in the Waikato Region.* Hamilton: New Zealand Geographical Society.

———, 2000. *The Legacy of Ngatoroirangi: Maori Customary Use of Geothermal Resources.* Hamilton: Department of Geography, University of Waikato.

———, 2002. *Wiremu Tamihana: Rangatira.* Wellington: Huia Publishers.

———, 2004. *Ohaaki: A Power Station on Maori Land.* Hamilton: Te Maataahauariki.

Strang, V., 2004. *The Meaning of Water.* Oxford: Berg.

Tahana, Y., 2011. 'Iwi Leaders Talk Asset Sales with PM.' *New Zealand Herald,* 17 February.

Te Hurinui, P., 1959. *King Potatau: An Account of the Life of Potatau Te Wherowhero the First Maori King.* Wellington: The Polynesian Society.

Titchall, A., 2008. 'The River Room.' *Energy NZ* 4 (Autumn).

Waitangi Tribunal, 1993. *The Pouakani Report.* Wellington: Brookers & Friend IV (Waitangi Tribunal Report 0113–4124: 6WTR).

Walker, R., 2004 [1990]. *Ka Whawhai Tonu Matou, Struggle Without End.* Auckland: Penguin Books.

7. Waters of Destruction: Mythical Creatures, Boiling Pots and Tourist Encounters at Wailuku River in Hilo, Hawai'i

EILIN HOLTAN TORGERSEN

Introduction

The relationships between people and the sea, or the 'social life' of the sea, have been extensively studied in the Pacific, yet the social lives of rivers in this region have not been given the same attention. Given the importance that rivers hold for most societies, it makes sense to ask whether rivers in the Pacific are equally as interesting and complex as the sea. Looking at large rivers on a global scale, such as the Ganges, the Amazon or the Nile—all rather striking natural features that occupy large areas within the landscape—one becomes aware of the important roles such rivers have played in world history and in the development of the societies along their banks. These rivers are roads for transportation and trade, pilgrim destinations, sources for irrigation systems and drinking water, tourist attractions and, in general, gathering places for people of different social and cultural backgrounds.

Like others in this volume (Chapters 3, 4 and 9), this chapter will focus on bodies of fresh water on a much smaller scale. Hawaiʻi does not have any large-scale rivers. In fact, many of the rivers in Hawaiʻi would probably be called creeks or streams in societies with large-scale rivers, or even those with mid-sized rivers that are discussed in this volume (e.g. Chapters 2 and 5). So why should we study rivers in Hawaiʻi? The small scale of the rivers does not make them less important or less interesting; on the contrary, they prove to be equally as important to the local landscape, economy, culture and history as any large-scale river elsewhere. Long before Cook arrived in the islands in 1778, for instance, Hawaiians had developed irrigation systems by digging ditches (ʻauwai) and constructing flumes that directed water from rivers and streams to their gardens and to larger crop fields such as those used in the cultivation of taro (McGregor 2006: 49–50). According to MacLennan (2007: 497) water was also involved in politics and spirituality:

> Water regulation ordered much of the agricultural and political social system in the taro producing valleys of Hawaiʻi. Deeply integrated in the political hierarchy and spiritual world, [water] was a foundation of life and meaning. Water was to be shared so that life was possible. It was water, rather than land, that formed one of the central cornerstones of life.

Miike (2004: 56) claims that:

> Water symbolized bountifulness, because it irrigated taro, the Hawaiian staff of life. Taro loʻi [a patch of wetland dedicated to growing taro] were irrigated by diverting stream water through ʻauwai or ditches. Water was of such importance to Hawaiian agriculture that stated twice—waiwai—it was the word for 'wealth'. The equal sharing of water—kānāwai—also represented the law: a person's right to enjoy his privileges and conceding the same right to his fellow man.

The Great Mahele of 1848, a piece of legislation that took the first step towards privatising land in Hawaiʻi, brought forth the first change in the meaning of, and access to, water resources (Chinen 1958). When the sugar industry started to boom in Hawaiʻi, beginning in the 1870s with a push by Euro-Americans towards industrialising sugar production, many of the water sources were transformed from shared resources (under Hawaiian law) into resources controlled mostly by the sugar companies (MacLennan

2007: 497).[1] Today, irrigation for farmland still comes from the streams and rivers in the islands, and represents one of many important factors in the relationships between people and rivers.

Water and rivers in contemporary Hawai'i hold many meanings and are subject to multiple contestations. As Strang (2004: 1) argues in a global context:

> All over the blue planet, even in the most rained-upon nations, people are engaged in conflicts over water. There are debates about who should own it, manage it, have access to it, profit from it, control it or regulate it. Nothing on earth, not even land, is more contested.

Following Strang, this chapter will examine who has or should have access to and control over a particular river in Hilo on Hawai'i Island (also referred to as the Big Island). While presenting the river's features and how the river is utilised in contemporary Hilo, I also focus on encounters between tourists and local people,[2] as well as encounters between the latter and the river itself. The chapter is organised in three sections: the first section describes some of the traditional meanings of water and rivers in Hawai'i; the second provides an ethnographic and historical description of the Wailuku River on Hawai'i Island; and the third section looks at social encounters along this particular river.

Through participant observation I have subtly been made aware of the fascinating relationships and everyday interactions between people and their natural surroundings, including rivers, in contemporary Hawai'i. When reading through various anthropological writings on nature and land in Hawai'i, I have found that much of the literature emphasises property

---

1   See also Wilcox (1996) and Miike (2004) for more information on these topics.

2   The term 'local' is broad and often difficult to define in Hawai'i. Discussions about what it means to be local in Hilo have been continuous throughout my decade of doing research there, and I suspect for far longer than that. Every now and then, and also quite recently (during the autumn months of 2016), local and national newspapers have published feature articles, often written by people who defend their own local identities, which discuss what it means to be local, or the differences between being a local, Indigenous Hawaiian or being *haole*, another complicated term referring to people of Caucasian descent. In an article in the *Huffington Post* (Riker 2015), Ty Kawika Tengan, an associate professor of anthropology and ethnic studies at the University of Hawai'i at Mānoa, stated that 'I think [being local] means spending a significant amount of time in the islands so you're rooted in the community … A sense of localness is one that doesn't erase Native Hawaiian history … I see local … as how invested they are at maintaining Hawaii as a unique place'. I will use the term 'local' in this chapter in much the same way as Tengan uses it here, not pointing towards a specific ethnic background but more towards how people relate to the community and the history of Hawai'i.

rights in land, the development of land and what the relationship between land and people was like in pre-colonial times—all of which is usually tied to a political cause or form of analysis. While these are all important issues, such approaches fail to deal with the way that the contemporary everyday relationships between people and their natural surroundings are being played out.[3] Local residents, especially non-Indigenous residents, do not generally practice traditional reciprocal relationships with land and water, but they do construct intimate relationships with various features of the local environment. People in Hilo often use their natural surroundings as a playground, as exemplified by kids jumping into the river or playing in the ocean, teens and adults camping and hiking in the valleys and beaches, or even snowboarders riding through the first snow on Mauna Kea, the highest mountain in Hawai'i. How do people move about in their environment and how are notions of agency in people and land played out? How can stories about a river give us indications of social relationships and cultural practices? As for the Wailuku River itself, it has been poorly studied by anthropologists. The most extensive publications on this river lie within the fields of geology and archaeology, and not much has been written on its social life.

Water and Rivers in Hawai'i

Just like the people of the Gambier Islands (Mawyer, this volume), Hawaiians distinguish between salt or sea water (*kai*) and fresh water (*wai*) in their vernacular language. In Hawaiian mythology, fresh water is often associated with the deity Kāne in his capacity to bring forth water when the land is dry. According to Miike (2004: 45), the Hawaiian explanation for the origin of springs is that *wai* was female, and would gush out when Kāne, 'the male procreative force, thrust his spear into the ground'. Salt water is often used in cleansing rituals (Beckwith 1970), so a *pī kai* is 'a ritual sprinkling with sea water or other salted water to purify an area or person from spiritual contamination and remove *kapus* (taboos) and harmful influences' (Pukui et al. 1972: 179). This ritual is today often performed within the *hula*[4] tradition, and especially within *hula kahiko*

---

3   See McGregor (2006) for an exception to this approach.

4   *Hula* is a comprehensive tradition that involves knowledge about Hawaiian genealogies, history, cosmology, geography, ritual and ecology. The tradition is often expressed and performed through several different forms of dance, of which the two most common are *kāhiko* ('ancient') and *'auana* ('modern').

(ancient hula), in which the *kumu hula* (mentor/master of hula) sprinkles salt water in the corners and along the walls of the *halau* (the space for *hula* training) in order for her or his students not to be corrupted by unwanted spirits while in training.

The *hi'uwai* is a ritual that has much the same purpose as the *pī kai* but requires that a person's entire body be submerged in water. According to Pukui and Elbert's Hawaiian dictionary, *hi'uwai* refers to water purification festivities held on the second night of the month of Welehu (near the end of the year). 'The people bathed and frolicked in the sea or stream after midnight, then put on their finest tapa and ornaments for feasting and games' (Pukui and Elbert 1986: 72). Jacka (this volume) refers to the cleansing attributes of water in his story about Jonah, but whereas it seems that the water in this case purified or cleansed the physical body, a *hi'uwai* is a ritual focused on purifying the spirit or the mind. This ritual is often guided by a *kahuna* (priest or spiritual mentor), and can be performed in both salt and fresh water. My friend 'Alamea'[5] has written the following account of a *hi'uwai* as experienced during her renaming ceremony:

> I was born Piilani Kawailehua (climbing to the heavens on the raindrop of a *lehua* blossom) [last named removed],[6] named by my [paternal] grandmother Annie. In second grade, age 8, my mom had it legally changed to [first and middle name removed] (the rays of the sun entwined in the necklace of love) [last name removed]. She told me that I needed to change my name because I was having nightmares about dying. I used to dream that I was stuck in a raindrop that in essence was a drop of blood because of the tinge of the red *lehua* in the raindrop, and was dying, and I would start to ascend, wherever; heaven, paradise? I don't know but wherever you go after you die, the direction was up. This scared me when I was little and my mom too so we went and saw a *kahuna* [spiritual mentor] and he asked about my name and who named me. First he said that Piilani alone is a heavy name to carry for anyone as it is a name of royalty from the island of Maui. That is why Maui is known as Maui a Piilani (Maui of Piilani) and Big Island is known as Hawaii Moku o Keawe (Hawai'i Island of Keawe). Both Piilani and Keawe were the last rulers of both islands before Kamehameha united them. So in essence the best people to have these names are those from that family line and royalty. Second the correlation between the dream and my name was undeniable and he felt that the best way to stop the dreams was to rename me, and they needed to be stopped, because in Hawaiian culture many

---

5   This is a pseudonym.
6   Names have been removed in order to maintain the confidentiality of the interview participant.

times your dream is more than a dream, it can actually be an out of body experience. So, none of us wanted that and we decided to do a rebirth ritual. The process of renaming me went with me going to Ka`u with my mom to do a *hi'uwai* with the *kahuna* and her. A *hi'uwai* is a cleansing and rebirth ceremony where the person who is being reborn is dressed in a *pareo* [wrap-around skirt] and enters a river or the ocean (I was in the ocean) and is prayed over by the *kahuna* to cleanse and take away all bad *mana* and in my case the name not meant for me. I was then completely submerged under water three times (much like a baptism) and upon resurfacing the final time was reborn but unnamed. You then wait for a sign for a name. Luckily my sign came in the form of an *'iwa* [frigate] bird which was flying overhead in the sky. The sky was completely overcast and there wasn't a ray of sunlight anywhere but consequently, as we watched the bird, the clouds formed a ring called a *wana* in Hawaiian where the sun's rays shone through. I remember being cold and then when that happened just feeling so full of warmth and my mom and the *kahuna* did too and he told us that the feeling was more than just the heat and warmth from the sun but it was the feeling of love. As you can see my name has love twice in it and it represents the love we felt that day from the gods and the love a mother feels for her child and likewise. (Alamea, 6 February 2009)

While studying *hula* in Hilo in 2009, my *kumu* guided me through a *hi'uwai*, very similar to the ritual just described and also performed in the sea, in order to cleanse my soul of lingering spirits after having experienced a death within the family. Although many of the old traditions, rituals and myths connected to water are not practiced or recalled today, the ones just mentioned still have a significant relevance in the lives of contemporary Hawaiians. Water clearly still has a cleansing effect that allows Hawaiians to move away from haunted dreams and lingering spirits. However, the waters of Hawai'i are also believed to carry more ominous qualities, such as mythical creatures that can reside in and control the rivers and sea, and the Wailuku River is rumoured to be the resting place of such a creature.

Wailuku—Waters of Destruction

The Wailuku River descends from the upper reaches of Mauna Kea, Hawai'i's tallest volcano, towards the northwestern part of downtown Hilo on the Big Island and finally into Hilo Bay (see Figure 7.1). The river is 45 km in length, is situated within the *ahupua'a* of Pueo,[7] and is fed by many smaller streams such as Ho'okelekele, Kahoama, Pakahuahine and several others. According to the United States Geological Survey (USGS 2000), the river basin we see today was formed by at least two lava flows from Mauna Kea, the oldest being the 'Anuenue flow (10,500 years old), which formed the thick lip of Rainbow Falls, as well as most of the rounded grey boulders that can be found in another section of the river called Boiling Pots. The river channel was again filled by a *pahoehoe* flow about 3,100 years ago, which explains the mix of *a'a* and *pahoehoe* we see in the river channel today.[8]

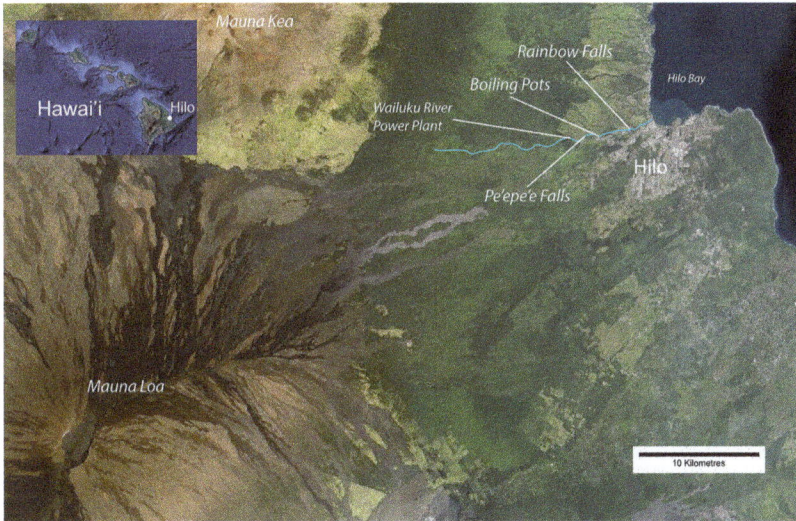

Figure 7.1 The Wailuku River.
Source: Cartography by Jerry Jacka.

Fed by rainwater, the Wailuku River flows, old and wise, fast and slow, with agendas often hidden from the untrained eye. While it can constitute an idyllic and somewhat romantic natural scene on calm sunny days, it can transform into a raging brown monster in the rainy season, boiling with

---

7   According to Pukui and Elbert (1986), *ahu* is a Hawaiian word for a heap, pile or collection, and *pua'a* is the word for pig. An *ahupua'a* is a land division, usually extending from the mountains or uplands to the sea, and so called because its borders were marked with heaps or piles (*ahu*) of stone, with an image of a pig (*pua'a*) or an actual pig as payment of tax to the chief placed on top of the pile.
8   Lava flows are divided into two main categories in the Hawaiian language—*a'a* and *pahoehoe*. An *a'a* flow is dark brown or black in colour and has sharp rock formations, while a *pahoehoe* flow is smoother and looks grey when placed next to an *a'a* flow in the landscape. These categories are used internationally within fields such as geology and volcanology.

fury as it thunders down towards Hilo Bay. According to the geological survey (USGS 2000), the river carries an average of 1 million cubic metres of water from its upper reaches all the way down to the Singing Bridge,[9] and in very rainy periods or during intense storms the discharge can be more than 20 times greater. It also transports up to 10 tonnes of suspended sediment into Hilo Bay. Because of the Hilo breakwater, which was built to protect the town from winter storms in 1929, the sediments remain trapped within the bay, prompting the rumour that its infamous brown murky waters are infested with reef and tiger sharks. Some of the water carried by the Wailuku River is turned into power for Big Island residents by several hydro-power stations located in the lower reaches of the river. The Hawaiian Electric Company (HELCO) built its first hydro-power station at Pʻuʻueo in 1910 and a second station at Waiau in 1920. In 1993, the Wailuku River Hydroelectric Power Company plant, located just below the junction of the Wailuku River and the Kalohewahewa/Hoʻokelekele and Kahoama streams, began commercial operation, providing additional power to HELCO. The hydro-power stations in the Wailuku River are a part of a larger effort to develop technology for renewable energy in the Hawaiian Islands. Even though the Wailuku River is a small river, not big enough or suitable for damming to create a more controlled level of power production, it plays an important part in the larger effort to 'go greener' in Hawaiʻi; thus, for some Hawaiians, it represents a green power alternative responsive to wider global warming and climate change debates.

In addition to serving as a source for power in Hawaiʻi, the Wailuku River also serves as a major tourist attraction in Hilo, especially at Rainbow Falls (called Waianuenue in Hawaiian), a beautiful waterfall shaped in the form of a wishbone falling into a pool of water in front of a big cave, at the Boiling Pots, a cluster of deep pools that 'boil' when the river is carrying large quantities of water, and at Peʻepeʻe Falls, a waterfall situated a little higher up the river from Boiling Pots (Figure 7.2).

---

9   The Singing Bridge is the last human-made construction associated with the Wailuku River before it flows into Hilo Bay. It is situated right on the shore line as one exits Hilo town on Bayfront Highway, and was given its name because it 'sings' when one drives over its metal grid base.

Figure 7.2 Waianuenue/Rainbow Falls.
Source: Photograph by the author.

Rainbow Falls and Boiling Pots have had viewpoint constructions added to their flanks, with parking lots, restrooms, easy-access ramps and security fences. Most tour operators in Hilo take their tourists by coach to see Rainbow Falls, and several take them up to Boiling Pots as well. Several helicopter tour companies offer tours of the upper reaches of the river that cannot easily be accessed by any other means of transportation, and from the air tourists can see the river's many upper waterfalls as well as trails leading through the Hilo Forest Reserve. At Rainbow Falls, visitors can see a large cave situated right behind the falls and, according to legend—as told by Mary Kawena Pukui—this cave was once the home of the demigod Maui's mother, Hina:

> Hina, mother of Maui, lives in a cave by the Wailuku River in Hilo on Hawaii where she beats bark cloth. While Maui is away at Aleha-ka-la (now called Hale-a-ka-la) snaring the sun, Lono-kaheo (some say Kuna the eel) comes to woo her and when she refuses him he almost drowns her. She calls to Maui for help and he throws about Lono-kaeho the snares with which he has overcome the sun and turns him into a rock which stands there today. The stone image of Hina could in the old days be seen with water dripping from its breasts, but a landslide has covered it. (Beckwith 1970: 232)

Lono-kaheo is a lizard-like creature that Hawaiians call a *moʻo*, a creature that can live in and control both salt and fresh water in the Hawaiian Islands. The living legend of the *moʻo* of Wailuku River claims that Lono-kaheo had its final resting place deep down in the dark pools of Boiling Pots, where it waits patiently to grab hold of and drag its next victim down into hidden lava tubes. Wailuku literally means 'waters of destruction'; while there is no doubt in the minds of the people of Hilo that Wailuku can be dangerous, it is still continually disturbed by reckless visitors and challenged to uphold its reputation of being a wrathful river that consumes its victims at will. Young Hilo residents swim in the river, especially at Boiling Pots, where they dare each other to jump from pot to pot down the river trail, even when the river is rough. According to local knowledge, someone dies here almost on a yearly basis, swallowed by currents caused by underground lava tubes in the deep pools, trapping the victims in pockets of water under the surface with no chance of getting out. According to Clark (2001), residents on Reeds Island, located downstream from Rainbow Falls, made several efforts to restrict access to the river after a long run of drownings of very young local swimmers in the mid to late 1990s. It was not possible to enforce these restrictions, however, since many of the popular spots where young people swim, such as 'the slides', 'the ropes' and 'South America', can be accessed through private properties that run all the way down to the water's edge. When fire service rescue divers were asked about how many people drown in the river, they said that no one has the exact numbers, but they would estimate it to be about one life claimed per year. 'Swimmers in distress in the Wailuku seldom survive', said Kala Mossman, who had been a rescue man for 10 years and had 17 years of scuba diving experience. That's because the 'currents are so strong, and the swimmers often are young or inexperienced' (Clark 2001).

After a period of heavy rain in early April 2014, I decided to go up the river to Boiling Pots to see if there would be anyone swimming there, and sure enough I found a group of local teenage boys diving into the boiling brown water and daring each other to make acrobatic jumps from the cliffs. One of them was standing on the riverbank watching his friends jump, and I went up to him to ask him why he was not jumping into the river with his friends. He said that he would never jump in, and stated: 'Nah, they're crazy, it's really dangerous when it's like this! People die here you know'. I asked him why he thought his friends would jump in and whether they did not know it was dangerous, and he said: 'Oh, they know, they just don't care!' I asked: 'Do you worry about them when they jump in the river when it's like this?' 'No,' he said, 'they just wanna have fun you know, it's all just fun'. 'So is it more fun to come here and swim when it's kinda dangerous then?', I asked. 'It is for them! I would never, I'm not allowed by my mom. But I wouldn't do it anyways, it's crazy, haha!' The boys let me hang out with them for a while, as I took pictures of their daring dives and expressed my concern for their well-being, before they moved towards Pe'epe'e Falls further up the river (Figure 7.3).

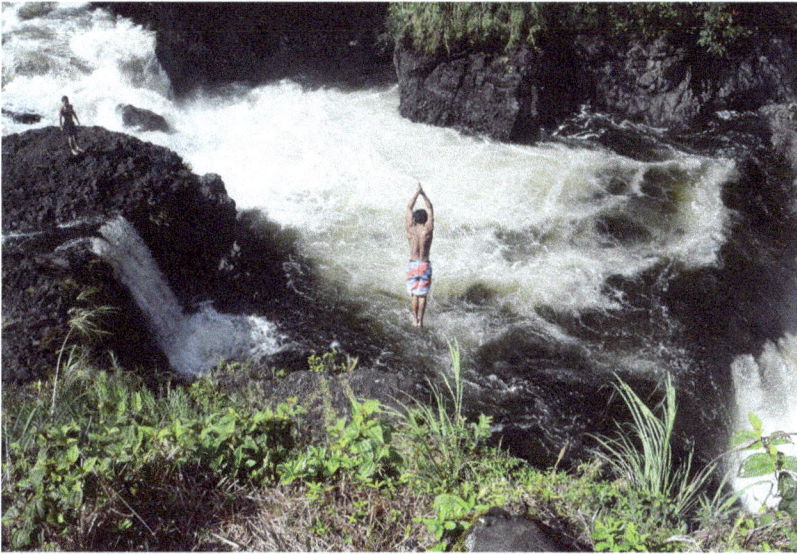

Figure 7.3 Local boy dives into a 'boiling pot' in Wailuku River.
Source: Photograph by the author.

In the aftermath of Hurricane Iselle, which hit Hawai'i Island on 7 August 2014, the Boiling Pots and Rainbow Falls areas of Wailuku River State Park were closed to the public, and guarded by security personnel, in order to keep people at a safe distance from the river, which at this point carried enormous quantities of water. Curious about what the river would look like after the hurricane, I went up to a section above Boiling Pots where two bridges cross the river. While approaching the bridges, I noticed cars parked alongside the road; when reaching the first bridge I realised I was not the only curious person who had set out to see the aftermath of the storm. Hilo residents had gathered on the bridge where they looked in awe upon the gushing brown water speeding down towards Hilo Bay and shared stories of what had happened to them during the storm (Figure 7.4). Everyone was making comments about how powerful the river was and that it would be 'no good swimming in it right now'.

Figure 7.4 People gather at Wailuku River after Hurricane Iselle.
Source: Photograph by the author.

While living in Hilo as an international exchange student for a year in 2007, I went swimming in Wailuku River at least once a week so long as the river was quiet. At this stage, I had not yet heard about the dangers and hidden agendas of this river. Sometimes we would bring masks and snorkels to explore the subsurface world of the river, revealing truly beautiful features such as large rock formations, archways and small caves formed by the lava flows that created the river base thousands of years ago. It is not difficult to understand why people are willing to partake in risky behaviour to experience the serenity and cooling effects of the river. However, as soon as I started doing *hula* in 2009, I stopped swimming in the river as my *kumu* preferred that I keep away from it.

Located just upstream from Rainbow Falls is the Hilo hospital, and after its construction the river has been said to be transporting toxic waste down to Hilo Bay. My *kumu* always said that if I absolutely could not stay away from the river, I should 'at least go swim above the hospital because of the waste'. 'And to think', she continued, 'what kind of toxic influence the *mo'o* is under, affecting its judgment in the acquiring of new victims. No, better you stay out altogether.' The belief that the water flowing in the reaches below the hospital is contaminated with disease, or filthy with toxic waste, is not necessarily rooted in the reality of the hospital's waste management, but there is a common belief that the location of the hospital has made the river's lower reaches unclean.[10] In addition to the contamination from the hospital, rivers and water in Hawai'i can often contain leptospirosis, a bacterial disease often transferred to humans via water containing urine from infected animals. People in Hilo are aware of this disease and claim that the Wailuku River contains both leptospirosis and E. coli. They would not drink from any river in the islands without purifying it with a filter or tablets first, and they say that you can never know what is happening upstream: 'A dead pig may be lying in the river up there'.

In spite of Wailuku's beauty and serenity on warm and sunny days, as well as its roles in the state economy as a tourist attraction and a source of renewable energy, it is the danger of the river that is mostly portrayed when people tell stories about it. A certain darkness rests over this river, and this is felt by most local residents because many have lost family members or friends to the Wailuku *mo'o*.

Tourism and Social Encounters in Wailuku River State Park

The rivers of Hawai'i are popular destinations for people visiting the islands. The picturesque waterfalls in these rivers are heavily romanticised by the tourist industry, and are experienced as highly exotic and sometimes even magical by visitors. Tourists come to Hawai'i searching for the perfect waterfall that dives into a refreshing freshwater pool, surrounded by lush vegetation and beautiful flowers—a perfect scene for romance or adventurous cliff-diving activities. However, their dream of paradise is often shattered when the local population or their tour guide warns them about dangerous currents, sharp subsurface rocks and what a pounding 50-metre-high waterfall will do to your head if you stand under it. Every year tourists run into problems while exploring areas outside resorts, and they often need assistance from lifeguards or search and rescue teams.

---

10 As Jacka writes (this volume): 'I never saw a Porgeran drink from a river. People are unsure of what others upstream may have done to or in the water.'

> First-time visitors to Hawai'i on packaged tours often arrive at Honolulu airport, are greeted with *leis*[11] and music by dark-skinned young people, and are bused to lavish hotels for a week of fun in the sun. They learn about something called the '*aloha* spirit' while on bus tours and absorb the friendliness of the Islands by observing smiling faces and receiving courteous, seemingly deferential treatment. All too many live in this plastic bubble, unaware that there is a complex society behind what may appear to be a gentle facade. (Haas 1998: 3)

While many areas in the Hawaiian Islands are very tourist friendly, other areas are considered off-limits to tourists. One example would be the Waianae coast on Oahu, which is rumoured to be the home of a community very hostile to tourists and newcomers. While taking a holiday with an Indigenous Hawaiian family in Waianae in 2009, I was made aware by neighbours and friends of the family that I was the only *haole* within a radius of several miles, and the neighbourhood rarely had visitors who were not local. I was told that I would not have been welcomed had I not been introduced to the community by their good neighbour. The *Honolulu Star Bulletin* published a story in 2004 about how the real estate market in Hawai'i has encouraged more continental Americans to buy properties in Waianae, and how these newcomers believe that the anti-*haole* image is a largely urban myth perpetuated by residents to protect the area. Larry Dingus, newcomer, resident and landowner in Waianae, was quoted as saying: 'I think there's a lot of self-interest in that image … . They don't want it to be "Californicated". It's a perception I'm happy to help keep alive' (Martin 2004). This concern about 'Californication' springs from an upsurge in the movement of people from California to Hawai'i, and is not limited to the Waianae area. I have often been made aware of this concern by residents in and around Hilo, especially in relation to places that are considered quintessentially Hawaiian, and these are often places that have great historical and mythical importance.

The complexity of relationships between people of different social backgrounds is not limited to that between residents and tourists in Hawai'i. The island group has been subjected to a continuous flow of migrants from different parts of the world, which first peaked with the sugar boom in the 1870s, when migrant workers were imported from Europe, Asia and the Pacific to work on the sugar plantations. Today,

---

11   A *lei* is an ornament worn around the head or about the neck. It can be made from flowers, leaves, shells, nuts, beads or feathers.

descendants with mixed ancestry from these migrants often identify as some form of 'local', often with a more specific twist to the term such as 'local Filipino', 'local Portuguese' or 'local Japanese'. As mentioned earlier in this chapter, people of Caucasian decent are usually referred to as *haole* and, even if they were born in the islands, it is difficult for them to identify as local. However, the term 'local *haole*' is used by some, and seems to be increasingly accepted as the debate on what it means to be local continues in Hawai'i. Other large social groups include the Japanese, Chinese, Micronesians and Indigenous or native Hawaiians known as Kanaka Maoli.[12] Relationships between these social groups often involve the negotiation of identity and sometimes give rise to tension, and they are part of the 'complex society' that Haas (1998) argues tourists do not experience. Even so, the concerns about a changing Hawai'i mentioned earlier may sometimes lead to uncomfortable encounters between residents and tourists.

Since Rainbow Falls and Boiling Pots are popular hangout spots frequented by young Hilo residents, as well as being popular sites for tourists, the Wailuku River represents a good example of a Hawaiian place for encounters between the local population and tourists. Sometimes these encounters can be rather tense, and the parties involved can be left with negative experiences. For instance, the *Honolulu Star Advertiser* reported on 6 February 2013 that a 49-year-old male visitor from China became separated from a larger tour group, and was then assaulted and robbed of his belongings near the restroom area at Rainbow Falls (Anon. 2013). While Rainbow Falls State Park seems—and usually is—a safe and serene place to visit, occasionally there are incidents like this that could threaten tourists' desire to make the effort to come and see it.

I witnessed another example of a negative encounter between visitors and the local population when doing fieldwork in Hilo in 2009. I was in Rainbow Falls State Park to look for banyan leaves for making *lei* for my *hula* class. Alongside the waterfall and the river is a path that leads down

---

12 The (problematic) contemporary legal definition of 'native Hawaiian' states that: '[a] Native Hawaiian [is] a descendant with at least one-half blood quantum of individuals inhabiting the Hawaiian Islands prior to 1778' (Kauanui 2008: 2). This definition originates from the *Hawaiian Homes Commission Act of 1921*, but is now surrounded by additional complications (Kauanui 2008; Kelley 2008; Torgersen 2010). The struggle for Indigenous Hawaiian rights has been going on since the early 1970s, with conflicts over land use and resources, culture, education and social benefits (Kanahele 1979; Linnekin 1983; Trask 1987, 1999, 2000; Kame'eleihiwa 1992; Friedman 1993; Silva 2004; Kauanui 2008; Kelley 2008; Tengan 2008; Torgersen 2010).

to a large banyan tree. In front of the waterfall is a paved viewpoint that stretches down to a parking lot for cars and buses, with restroom facilities at the back of the lot. The largest bus companies and tour operators on the island offer tours to see the waterfall and surrounding areas, and on this particular day a group of older tourists had made their way up to the park. While walking back down the path to the parking lot, I heard the squealing of car tyres and saw a car speeding through it, barely missing the elderly people making their way across the parking lot towards the viewpoint. A local girl was hanging out of the window of the car screaming, with the utmost power of her voice, 'Fucking haoles! Go back to where you fucking came from! We hate you and don't want you here! Fuck you haoles!' Obviously in rage, the local girl expressed in words and action what many locals feel about tourists, and her protest was likely rooted in frustration over their constant presence at Rainbow Falls.

However, such encounters of tension, and in this case even rage, between locals and tourists are not the norm. Most encounters are friendly and welcoming on both sides, evident in the good reputation the people of 'the Aloha State' hold both nationally and internationally. Nevertheless, when tourists move outside the typical tourist zones, they sometimes encounter local and Indigenous Hawaiians who are not so eager to have them around. One of the reasons encounters can have a very complicated back story is the interdependent relationship between residents and tourists, since the island economy, including everything from large resorts to small local businesses, is so dependent on a certain level of tourism activity, and the tourists themselves are dependent on the residents working within the industry.

Because of this interdependent relationship with tourists, locals and Indigenous Hawaiians have developed a certain expertise in impression management (Goffman 1959), and either as individuals or groups have adapted to lives in which they practice behavioural alteration in order to deal with the more than 8 million tourists who visit the islands every year (HTA 2012). This even exceeds the typical performativity found in most tourist/local encounters, where the tourist industry worker performs a role suited to the encounter. This sort of performance exists so long as the performer repeats the acts that signify it, as the social relationships are continually made out of practices, rather than prescriptively structured (Sahlins 1985: xii). Moving beyond the classical performances of tour guides or resort bartenders, for example, Hawaiians have cleverly developed methods of performance and impression management that keep tourists away from certain areas, or at the very least encourage them to interact more respectfully at places that hold special meaning within Hawaiian culture. The stories of the *mo'o* in Wailuku River are told in order to keep people away from the river, both because of the danger that it poses but also, I would suggest, to control the movement of people in and around the river.

Hawaiian mythology and cosmology are often used in stories told to tourists about certain places. On Hawai'i Island, tourists are told stories about the fire goddess Pele, and how her temperament should not be tested. They are told to act respectfully when entering and staying in areas that are most specifically considered to be Pele's land, or else she will punish them in one way or another. Another story, told especially to hikers and campers, concerns the Night Walkers, an army of the ghosts of Hawaiian warriors who march to old battlegrounds or spiritual places at sunrise and sunset. If you encounter the Night Walkers you must lie down with your face towards the ground to show them respect, or else, so the story goes, you will die.

Hviding (1998) argues that the people of Marovo Lagoon in Solomon Islands also practice impression management when they use their dark past as headhunters, as well as the dangers of their natural environment, in an instrumental way in order to uphold a form of spatial discipline and control the movement of the tourists they encounter. Just as people in Hilo would tell tourists stories about the Wailuku *mo'o* and Pele's unforgiving temperament, people of Marovo take their tourists on hikes into the jungle to see shrines containing actual human skulls from the headhunting period, or take them up murky rivers while telling scary stories of big man-eating crocodiles.

Hviding gives an account of a group of entrepreneurs (or adventure tourists) who came to Marovo to set up a kayaking business, but were caught off guard by the complex land tenure system in the lagoon, as well as by the constant presence of villagers, which shattered their illusion of 'that secluded spot [in which they] could lounge about undisturbed' (Hviding 1998: 1). As they paddled about the lagoon, looking for places to camp and places to enjoy the fruits of the land and sea, with a fresh memory of village stories about shark-infested waters and man-eating crocodiles, rain, wind and darkness caught up with them. While fearing the rough waters and what they could potentially be hiding, and after punching a hole through their kayak on the coral reef, the group started to panic, yet still managed to find a site where they 'silently set up camp' for the night and waited for the morning sun to rise. As soon as morning came, they headed straight for the nearest resort where they could enjoy good food, warm showers and comfortable beds.

> For the anthropologist with years of experience from Marovo Lagoon and close enough knowledge of people and place to be able to trace the movements and interactions recounted, the kayakers' story turns into a tale of uncertain encounters in which the much-praised pristine nature and friendly people of Marovo are attributed with dark secrets and adventure tourism becomes a demanding enterprise and potentially gruelling experience for the tourist. Sharks and crocodiles hungry for human flesh appear to lurk behind the increasingly shattered facade of a Paradise full of gusty winds and sharp coral waiting to rip the bottom out of any flimsy craft passing overhead. (Hviding 1998: 41)

The tourists came looking for Paradise, but ended up in what Hviding calls a 'fearsome Heart of Darkness'. Paradise could only be restored by escaping the scary unknown and entering a more comfortably constructed kind of paradise that represented something entirely different from the adventurous ecotourism the group had set out to experience.

In the Hawaiian case, scary stories about the *mo'o* living in the hidden caves and lava tubes in the Wailuku River turn a beautiful natural feature of the landscape into something ominous, thus keeping the hordes of tourists out of the river and away from a favourite spot that Hilo locals prefer to enjoy in seclusion. However mythical, the agency attributed to the *mo'o* structures the reality of the river, gives it a kind of agency beyond human control, and thus constitutes a kind of 'mythical reality' (Sahlins 1981).

Conclusion

I started this chapter by questioning why we should study rivers in the Pacific, or more specifically in Hawai'i. While briefly looking into the meaning and importance of water, this chapter has mainly focused on the role of the Wailuku River in contemporary Hilo society. With its powerful reputation as a destructive river, it is attributed agency by its visitors and considered to live a life of its own. Studying this river can give us insights into relationships between the residents of Hilo and their natural environment, as well as their relationships to the myths of old Hawai'i. The social life of this river involves relationships on different levels, since it is not only a source of 'clean power' for Big Island residents and irrigation for its surrounding farmland, but also the basis for a Big Island tourist industry. Particular sections of the river are also primary locations for encounters between the local population and visitors to Hilo, as they are popular spots for both tourist groups and local youth. The encounters exemplified in this chapter tell us that the relationships between the locals and the visitors can sometimes involve tension, even rage, and are often complicated because of their interdependency. While making room for the tourist industry, Hilo residents have also cleverly developed a way of enforcing a form of spatial discipline along the Wailuku River by telling stories about the river's cunning agency in order to gain some control over the movement of visitors. The local population can thus enjoy the river in some measure of seclusion, even while the *mo'o* patiently lies waiting for its next victim in the river's dark boiling pots.

References

Anon., 2013. 'Chinese Visitor Robbed at Rainbow Falls Park in Hilo.' *Honolulu Star Advertiser*, 6 February.

Beckwith, M., 1970. *Hawaiian Mythology*. Honolulu: University of Hawai'i Press.

Chinen, J.J., 1958. *The Great Māhele: Hawaii's Land Division of 1848—Volume 1.* Honolulu: University of Hawai'i Press.

Clark, H., 2001. 'Many Young Lives Lost in Hilo's Wailuku River.' *Honolulu Advertiser*, 15 June.

Friedman, J., 1993. 'Will the Real Hawaiian Please Stand: Anthropologists and Natives in the Global Struggle for Identity.' *Bijdragen tot de Taal-, Land- en Volkenkunde* 149: 737–767. doi.org/10.1163/22134379-90003111

Goffman, E., 1959. *The Presentation of Self in Everyday Life*. New York: Doubleday.

Haas, M., 1998. *Multicultural Hawai'i: The Fabric of a Multiethnic Society*. New York: Garland Publishing.

HTA (Hawai'i Tourism Authority), 2012. '2012 Annual Visitor Research Report.' Viewed 28 October 2013 at: files.hawaii.gov/dbedt/visitor/visitor-research/2012-annual-visitor.pdf

Hviding, E., 1998. 'Western Movements in Non-Western Worlds: Towards an Anthropology of Uncertain Encounters.' *Journal of the Finnish Anthropological Society* 23(3): 30–51.

Kameʻeleihiwa, L., 1992. *Native Land and Foreign Desires*. Honolulu: Bernice P. Bishop Museum Press.

Kanahele, G.S., 1979. 'The Hawaiian Renaissance.' Kamehameha Schools Archive. Viewed 3 April 2010 at: kapalama.ksbe.edu/archives/pvsa/primary%202/79%20kanahele/kanahele.htm

Kauanui, K.J., 2008. *Hawaiian Blood: Colonialism and the Politics of Sovereignty and Indigeneity*. Durham (NC): Duke University Press. doi.org/10.1215/9780822391494

Kelley, K.A., 2008. *Noho Hewa: The Wrongful Occupation of Hawaiʻi*. Independent documentary film.

Linnekin, J., 1983. 'Defining Tradition: Variations on the Hawaiian Identity.' *American Ethnologist* 10: 241–252. doi.org/10.1525/ae.1983.10.2.02a00020

MacLennan, C.A., 2007. 'Wai: Indigenous Water, Industrial Water in Hawaiʻi.' *Organization & Environment* 20: 497–505. doi.org/10.1177/1086026607309389

Martin, D., 2004. 'Changing the Face of Waianae: Mainland Buyers Are Injecting a New Element into Oahu's West Side.' *Honolulu Star Bulletin*, 16 May.

McGregor, D.P., 2006. *Nā Kuaʻāina: Living Hawaiian Culture*. Honolulu: University of Hawaiʻi Press.

Miike, L.H., 2004. *Water and the Law in Hawaiʻi*. Honolulu: University of Hawaiʻi Press.

Pukui, M.K. and S. Elbert, 1986. *Hawaiian Dictionary: Revised and Enlarged Edition*. Honolulu: University of Hawaiʻi Press.

Pukui, M.K., E.W. Haertig and C.A. Lee, 1972. *Nānā I Ke Kumu (Look to the Source)—Volume 1*. Honolulu: Hui Hānai.

Riker, M., 2015. 'What Does It Mean to Be Local in Hawaii?' *Huffington Post*, 2 October.

Sahlins, M., 1981. *Historical Metaphors and Mythical Realities: Structure in the Early History of the Sandwich Islands Kingdom*. Ann Arbor: University of Michigan Press. doi.org/10.3998/mpub.6773

——, 1985. *Islands of History*. Chicago: University of Chicago Press.

Silva, N.K., 2004. *Aloha Betrayed: Native Hawaiian Resistance to American Colonialism.* Durham (NC): Duke University Press. doi.org/10.1215/9780822386223

Strang, V., 2004. *The Meaning of Water.* Oxford: Berg.

Tengan, T.P.K., 2008. *Native Men Remade: Gender and Nation in Contemporary Hawai'i.* Durham (NC): Duke University Press. doi.org/10.1215/9780822389378

Torgersen, E.H., 2010. 'The Social Meanings of Hula: Hawaiian Traditions and Politicized Identities in Hilo.' Bergen: University of Bergen (MA thesis).

Trask, H.-K., 1987. 'The Birth of the Modern Hawaiian Movement: Kalama Valley, O'ahu.' *Hawaiian Journal of History* 21: 126–152.

——, 1999. *From a Native Daughter: Colonialism and Sovereignty in Hawai'i.* Honolulu: University of Hawai'i Press.

——, 2000. 'Natives and Anthropologists: The Colonial Struggle.' In D.L. Hanlon and G.M. White (eds), *Voyaging Through the Contemporary Pacific.* Lanham (MD): Rowman and Littlefield.

USGS (United States Geological Survey), 2000. 'The Wailuku River: Mo'o and Lava.' Viewed 20 March 2013 at: hvo.wr.usgs.gov/volcanowatch/archive/2000/00_06_22.html

Wilcox, C., 1996. *Sugar Water: Hawaii's Plantation Ditches.* Honolulu: University of Hawai'i Press.

8. The Sepik River, Papua New Guinea: Nourishing Tradition and Modern Catastrophe

ERIC K. SILVERMAN

… although, it brings so much to the people, they know it can never be trusted and they will never laugh at it. It is there for their use but also for their veneration. Too many of them have been sucked under by its temperamental currents and its grisly executioners, the crocodiles; too many of their dwellings and gardens have been ruined by its raging flood; too many canoes and boats have floundered on its submerged logs for men to regard it lightly. All its moods and changing scenes of mist shrouded rain squall, glistening rainbows, harsh sunshine, dawns and sunsets and delicate moonlight add to its peculiar mystique. (Unnamed writer, *Papua New Guinea Scene,* October 1970, quoted in Leigh and Perry 2011: 200)

Introduction

In 2010, I returned to Tambunum village, an Eastern Iatmul community along the banks of the Middle Sepik River in Papua New Guinea. Fifteen years had elapsed since my last fieldwork season. A lot of water had passed under the proverbial bridge—a *lot* of water. And lest I appear to impose an alien figure of speech on local experience, I note that Eastern Iatmul understand the Milky Way as a celestial bridge that spans the Sepik.

From Wewak, the commercial and political hub of East Sepik Province, my Iatmul companions and I boarded a passenger truck bound for the river. Hours later, we disembarked at the river's edge in Angoram, one of three administrative centres along the Sepik. We hauled our gear into a fibreglass dinghy equipped with a 40-horsepower outboard motor, a type of vessel increasingly seen nowadays, and headed upriver for about 80 km. As in years past, I glimpsed few signs of modernity. But one actually sees very little while travelling along the Sepik—mainly dugout canoes, tall reeds and swordgrass, the occasional hamlet and a few distant hills. For the most part, the river permits a view of only its own presence, a vast aquatic world endlessly in flow.

We travelled the main serpentine watercourse for several hours, often navigating slowly through smaller channels that slice across the many oxbow bends and lagoons. Then we rounded a familiar point. I would again, after so many years, 'be there' in the field and reunite with village kin. But as I clambered out of the dinghy that drizzling afternoon, my nostalgic elation turned to bewilderment. The village had largely vanished. In its place was water.

Globally, the Sepik is a relatively minor watercourse—ranked 145th in terms of its length (Wikipedia 2015a). Yet, the Sepik is often hailed as the largest 'pristine' river in the world, unsullied by bridges, dams, pollution or industry (Wikipedia 2015b). Global comparisons, however, fail to capture the enormity of the river in its local eco-cultural setting. Indeed, the region is not so much a landscape as a 'fluidscape' (Strang 2006: 149). The Sepik is a traditional mother figure of nurture and sustenance (Silverman 2001: 15). But the river also has an appetite, especially in the last few years, swallowing huge parcels of ground in Tambunum, including houses and trees (see Figure 8.1). During my most recent fieldwork in 2014, erosion dominated village conversations. The very 'ground' of social life is dissolving.

Figure 8.1 Where village land was, is now only river.
Source: Photo by the author.

From source to mouth, in a direct line 'as the crow flies', the Sepik covers a distance of just over 402 km. But the river flows for almost three times that distance, constantly twisting anew, relentlessly washing away any sense of stability. Local men do their best to anchor the world, but such gestures of permanence and solidity are illusory, especially today. Reality is fluid, both in terms of traditional cosmology and people's confrontation with modernity. Moreover, as the Sepik increasingly threatens the viability of the village, it also surges as a local metaphor for broader struggles amidst the lack of 'development'. And while the name 'Sepik', which dates to the colonial era, persists as an important post-colonial regional signifier, it also betokens a sense of backwardness, because the river is a kind of 'backwater'. My overall argument, then, focusing on recent flooding but also on more distant historical events, is that the Sepik River symbolises contrary or dialogical[1] meanings: future and past, land and water, prosperity and underdevelopment, male and female, culture and loss.

---

1   For the theory of cultural dialogics that frames this essay, see Lipset and Silverman (2005).

Torrential Water and Magical Dirt

During the 2009–10 rainy season (roughly December to April), Tambunum suffered a devastating flood. The river overflows every year, but this was altogether different. Gardens were inundated, fruit trees died, dogs and chickens drowned. Anything not lashed to a house post, or pulled inside a dwelling, floated away. Waves rolled through the village as if they came from the ocean. Paddling was impossible. 'The water was as dark as the sea', people told me. Women cooked indoors, and so house fires were a constant danger. Most people fled to a hill in the hinterlands after travelling downstream for a few hours.

I arrived in 2010 at the start of the dry season. By then the village was dry and the river had receded. But people spoke of a 'disaster'. Dozens of families had lost their land to the flood, and so took up residence elsewhere in the region. The community was dispersed, lacking any 'centre', its vibrancy diminished when compared to the 1980s. The Sepik is overwhelming us, people said, it is *daunim ples* ('destroying the village' in Tok Pisin). Many advocated relocating the *entire* village elsewhere as the only viable option for rescuing the community from the water.

The establishment of a new village is no trivial endeavour. The practical labours are substantial—clearing trees, rebuilding houses, and so forth. But men emphasise instead the numinous perils associated with relocating the senior cult house. They must unearth dirt from beneath the current central posts and rebury this soil under the new structure. This is no ordinary dirt, having originated in the uterine, mythical pit that gave birth to the world. The same dirt was shaped into the first humans and ferried from place to place during ancestral migrations. By re-interring this magical dirt, men and *not* women 'replant' the totality of 'ancestral law' (*ara* in the eastern dialect of the Iatmul vernacular).[2] Of course, men *only* need to enact this rite of reburial on account of riverine erosion. The Sepik, it is crucial to note, is generally coded as female. Land and river, then, form antiphonal voices in a wider, ongoing dialogue about gender.

---

2   People in Tambunum village are bilingual in Tok Pisin and the eastern dialect of the Iatmul language. Those who have attended school—a sadly decreasing number of children these days—can also converse in English to differing degrees. My fieldwork has made use of all three languages.

In the Indian state of Maharashtra, rivers represent abundance, nourishment and cleansing (Feldhaus 1995). Yet riverine fecundity is wild and perilous, like women's menstrual flows. The Sepik evokes similar meanings, at once a predictable source of everyday sustenance but a source of floods and erosion that humans cannot tame and that threaten the 'ground' of sustainability. The ancestral spirit of the Sepik will also punish—even slay—the kin of anybody who frolics in the middle of the river or dares to swim across. But Eastern Iatmul do not fear the water or its denizens, including crocodiles. Children learn to master the river, playing and swimming alongside moored canoes. Nevertheless, Eastern Iatmul perceive the Sepik as something 'wild', especially today, and even as an obstacle to 'development'. Other regions of Papua New Guinea (PNG) benefit from petroleum, natural gas, gold and copper. But the people of Tambunum say that they 'only have water and fish'. They also mention another resource, namely woodcarvings made for tourists, but there are few tourists nowadays. So distraught are some villagers with their current relationship to the river that they now advocate abandoning the Sepik entirely, thus reconfiguring their identity from being 'river people' to being 'bush people'. There could be no more striking an admission of defeat.

Currently, what the Eastern Iatmul refer to as an unnamed 'Asian' company is harvesting hardwood trees on land claimed by the Mogua (or Fish) clan, several hours downriver from the village. Until recently, the logging was allegedly illegal, or so the people of Tambunum say, concealed by a licence for palm oil. In April 2015, however, I received an email copy of a semi-completed 'customary landowner's consent' form, issued by the PNG Forest Authority, on which 75 village men approved the harvesting of timber by Summit Agriculture Limited. The company provided an address in Port Moresby, the capital of PNG, but a few minutes with Google revealed connections to Singapore and Malaysia. The consent form is nothing more than a legal fiction, declaring the timber otherwise 'wasted' from local people clearing the land for use as a 'garden area' (which, in fact, was never the case). In 2014, I saw hundreds of logs stacked along a recently graded dirt road that leads from the harvesting sites to Wewak. Many men from the Fish clan, employed by the loggers, camped with their kin beside the road in shelters, several kilometres away from the river. It is here that some people propose establishing a new village, one entirely bereft of the riverine resources that define the Iatmul *as* Iatmul (see Figure 8.2). Instead of tying the village to the riverbank, like a canoe, this effort would root the community in the 'bush'.

Figure 8.2 Woman and child fishing.
Source: Photo by the author.

Traditionally, each of the roughly two dozen Iatmul villages exchanged fish for processed sago starch with their inland neighbours, the Sawos (Gewertz 1983). Three such nearby hamlets once envied the affluence of Tambunum, which mainly arose from the riverboat tourism that peaked in the 1980s and 1990s. Today, however, Eastern Iatmul aspire to the wealth currently enjoyed by some of the nearby Sawos villages, whose people they still deride as filthy, uncouth and uneducated, but whose hardwood forests now produce logging revenues. The economic, but not the moral, topography of the region has thus been inverted: the river, once an asset, is now viewed by Eastern Iatmul as a liability, while the bush, a former source of impoverishment, is seen as a source of prosperity.

Despite this reversal, Eastern Iatmul continue to identify as 'river people', and the Sepik remains a ubiquitous cultural motif. Thus, wave patterns still decorate almost all of their material culture (Figure 8.3). The river endures as a principal schema for organising reality into a fluid state of female flow that men seek to stabilise through their labours, such as building houses. To be sure, local people now contemplate this aquatic metaphysics with heightened ambivalence. The inability to control the river indexes a wider powerlessness to shape the fate and prosperity of the community. But the river remains the foundation for their ontology.

Figure 8.3 House ornamented with wave patterns.
Source: Photo by the author.

An Aquatic Presence and Modern Identity

The Sepik catchment area encompasses some 78,000 km², and nowadays more than 400,000 people—more than 70,000 in the floodplain alone (Dudgeon and Smith 2006: 207). During the rainy season, the river may swell in places to a width of 30–70 km for a period of five months. The depth of the lower reaches of the river may extend to 35 m, but sandbars quickly form in the shallower middle and upper reaches, ensnaring drifting trees and branches that can swamp or shatter canoes. Furthermore, thunderstorms can churn surprisingly high and erratic waves. Some stretches of the river flow between sun-baked mud banks that can rise several metres above the normal waterline, but other stretches spread into impenetrable marshes.

The main Eastern Iatmul village of Tambunum—or what remains of it—curves for about 1.5 km around a large point jutting from the southern bank.[3] The lower (downstream) half of the village slopes gently to the water, and the upper half once did likewise, but now a steep embankment, formed by slabs of ground tumbling into the river, makes for a perilous climb down to the water (see Figure 8.1). In the late 1980s, the main village contained more than 130 houses, but only 77 dwellings remained in 2014. The entirety of what was the upper half of the village in the 1980s and 1990s, which included dozens of dwellings as well as the main cult house and a long ceremonial plaza, is now water. As noted earlier, erosion has forced most households to take up residence elsewhere, mainly on the other side of the river, extending downstream for several kilometres. Tambunum today is not so much a well-defined village as an ideal that loosely enmeshes detached, dispirited hamlets. Many people yearn to revive tourism, yet there is frankly no 'place' for tourists to visit.

In 2014, Eastern Iatmul people remarked in various ways on their current inability to sustain a 'real' village. The community is unkempt, as tall grass overgrows abandoned paths and plots. While the grass attracts snakes, it also represents an intrusion of nature on culture. With the death of the older 'big men' and the loss of their esoteric knowledge, the village lacks the discipline and force of 'traditional law'. Theft is increasing; young people smoke marijuana and drink 'homebrew'; gardens are meagre; and there are few of the coconut palms that constitute a local metaphor for intergenerational sociality. Lacking manpower and money, moreover, the community is unable to construct the massive domestic houses for which it was once renowned. In the 1980s, over a dozen such dwellings, unparalleled in the region, enthralled tourists. Today there are none. And if they did build such houses, men say, the structures would only topple into the river. As one man remarked with disgust, the community now 'looks like a swamp'.

The river dominates the poetics, prosaics and mundane particularities of everyday life. The river, too, distinguishes Iatmul (and other communities who live along the Sepik) from the rest of PNG. People throughout the province self-identify as 'Sepiks', an ethno-regional designation that dates to the early colonial era. The word 'Sipik' was first reported by Full (1909: 339) as one of two names for the watercourse—the other

---

3   On Google Earth, you can view a 2011 image of the village, centered at 4°11'05.68"S and 143°35'54.47"E.

being 'Abschima'—that were used by 'natives' living at the mouth of the river. A few years later, Schultze (1914) applied the term 'Sepik' to the entire watercourse, and this usage prevailed. Regardless of its European derivation, the term came to define a key coordinate of local identity. In Tambunum, the vernacular name for the river is Avatset, a compound of the words for 'bone' (*ava*) and 'lake' (*tset*). But Iatmul do not really designate themselves as 'Avatset people'; they are simply 'Sepiks'. Thus, men chisel the name into the woodcarvings made for tourists, while women weave riverine references into the baskets they sell to other Papua New Guineans in town markets. These references include 'Meri Sepik', 'Sepik Souls' and 'PS', the latter being an abbreviation in Tok Pisin for 'Pikinini Sepik' ('Sepik Child') (see Figure 8.4). This phrase, in the context of local idioms of motherhood (Silverman 2001), declares that Eastern Iatmul people are born of the river. The Sepik is their mother.

Figure 8.4 Women selling baskets with 'PS' and 'Sepik'.
Source: Photo by the author.

In the past, Iatmul villages were threatened not just by erosion and flooding, but by crocodiles as well. In 1935, for example, Sarah Chinnery, whose husband was both an anthropologist and Australian colonial official, travelled up the Sepik and reportedly learned from an expatriate trader that '[c]rocodiles get a lot of women who go fishing round the low, swampy parts of the river' (Fortune 1998: 162–3). More recently, *The National*, a PNG newspaper, reported in January 1995 that a man in Tambunum 'dived straight into the mouth of the crocodile, watched by his helpless wife'. Today, however, crocodiles are scarcely seen or feared. After decades of being hunted for their skins, coupled with the effects of wetlands degradation, Sepik crocodiles are now the beneficiaries of conservation projects (UNDP 2012). They do inhabit the smaller tributaries and swamps, they do occasionally attack people, and in the local cosmology, crocodile spirits (*wai-wainjiimot*) continue to preside over daily and ceremonial affairs. However, most of the crocodiles seen today are decorative images (see Figure 8.5).

Figure 8.5 Crocodile clock sold by a man from Tambunum to a tradestore in Wewak.

Source: Photo by the author.

In fact, the crocodile (*pukpuk* in Tok Pisin) serves as the riverine mascot of pan-Sepik identity. Crocodiles adorn the provincial flag, the One Kina coin, hotel bars, clothing and woodcarvings sold to tourists and shopkeepers. The World Wide Fund for Nature (WWF) sponsors a quasi-annual Sepik Crocodile Festival. In town, women sell colourful beaded bracelets that proclaim 'PS Pukpuk' (short for 'Pikinini Sepik Pukpuk' or 'Sepik Crocodile Child'). The Wewak Christian Bookstore displays souvenir crocodiles beside posters of Jesus. A crocodile mask, draped with the PNG flag, ornaments the Avis car rental agency in Wewak. In 1994, two men from Tambunum carved a crocodile slit-drum in the New Guinea Sculpture Garden at Stanford University (Silverman 2003). I have even seen crocodiles painted on a pair of trousers, accompanied by the phrase 'PS trust me' (Figure 8.6).

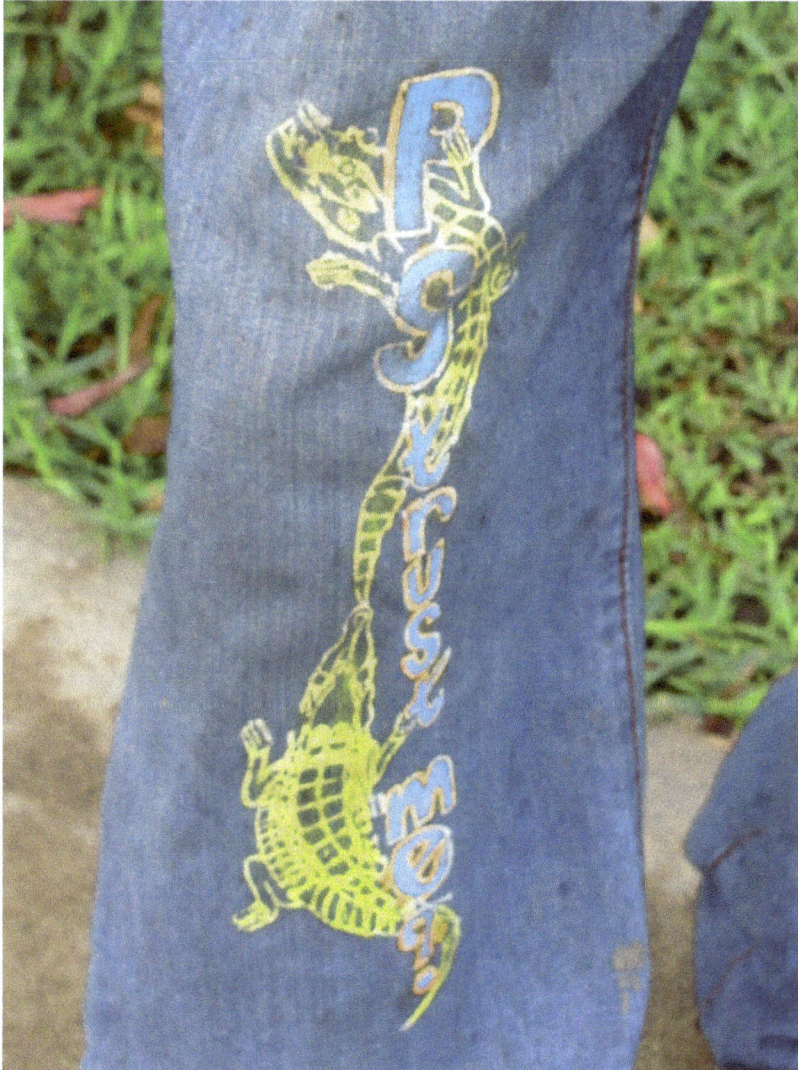

Figure 8.6 Decorated jeans: 'PS trust me'.
Source: Photo by the author.

The name 'Iatmul', like the ethnic signature 'Sepik', also dates back to the colonial era (Bateson 1932: 249n; also Claas 2009). But people from Tambunum never refer to themselves as 'Iatmul', even in contrast to other linguistic and cultural groups. I did hear a few utterances of the cognate 'Iatmoi' in the 1980s, but this usage seemed forced. It is fair to say that Iatmul villages and speakers simply do not unite *as* Iatmul. They do, however, identify today as 'Sepiks'. But in consequence of the dialogical role of the river in their everyday lives today, the people of Tambunum emotionally respond to their own self-identity with open ambivalence.

Historical Currents

In Iatmul myth, the beginning of the world was aquatic.[4] Wind stirred the primal water, and land surfaced amid the waves. A 'totemic pit' (*tsagi wangu*) cleaved the ground—just north of the middle reach of the river, near the Sawos village of Gaikarobi—and the ancestors emerged from it. They pushed up the sky and set out to create the 'paths' of the world. Ground materialised beneath their footsteps. Their descendants continued on these mythic journeys, often paddling canoes (see Figure 8.7), and named the features of the world as they came into existence. Each tale of creation forms a travelogue of land and water, the grand plot of which is the genesis of terrestrial differences from the original aquatic void.

The ancestors of the Shui Aimasa (Pig) clan created the land that spreads north of the river. The Mboey Nagusamay (Sago) clan claims the world that lies to the south. The Mogua (Fish) clan oversees the eastern world of the Lower Sepik, the ocean and everything overseas. And the riverbed—but not the water—is the totemic realm of a minor group called Wyngwenjap, whose ancestral eel and snake formed the riverbanks. Members of this clan are said to have filled the Sepik by pouring water from a magical bamboo container, given to them by a lineage of the Sago clan affiliated with the nearby Karawari River. The heads of the totemic eel and snake glower at each other, as if in a staring contest, at a narrow stretch of the upper river near Yambon village, featuring dangerous rapids and whirlpools. This was formerly the western limit of Eastern Iatmul knowledge. The tails extend beyond the mouth of the river to Manam,

---

4   The floodplain was covered by an inland sea during the late Quaternary period (Swadling and Hide 2005).

a volcanic island in the Bismarck Sea (Lutkehaus 1995). When the tails periodically touch, the Manam volcano erupts. The Sepik, then, served as a sort of *axis mundi* in Iatmul cosmology.

Figure 8.7 Paddling to school.
Source: Photo by the author.

Not surprisingly, the history of European contact with the Iatmul is a tale of the river. In 1884, Germany asserted colonial control over New Guinea. The colony was initially managed by the New Guinea Company, a commercial enterprise, which christened the territory Kaiser Wilhelmsland (Buschmann 2009: 35). The first European ship to enter the river's estuary was the Samoa in May 1885. The 'mighty' river, as it was described in the company's annual report for 1886–87 (Sack and Clark 1979: 15), was termed Kaiserin Augusta Fluss. A year later, the Samoa returned to the estuary and launched an open whaleboat in a failed effort to reach the Dutch–German border. The company and government officials returned some months later on a steamship to assess the economic potential of the river. Thereafter, German interest in the Sepik focused on collecting artefacts and native labourers, the latter being put to work on coastal copra plantations (Buschmann 2009: 105).

Of course, Europeans also travelled the Sepik to stake claims over native souls. In June 1887, the Samoa returned with another scientific expedition as well as a dozen Malays, eight men from New Britain and, most significantly, two members of the Rhenish Missionary Society. On 30 June, as the ship neared Tambunum, one of the missionaries wrote that 'everybody [in the village] ran back and forth shouting and grabbed for the spears' (Claas and Roscoe 2009: 340). Warriors gestured for the Europeans to move on. A few of the scientists set off in a boat to take photos. 'At that, the people got even wilder.' Much further upriver, local people fired arrows at the expedition, and the Europeans discharged their firearms in response. Following this expedition, the annual reports have nothing to say about the river for the next two decades.

But the Sepik was hardly forgotten, especially by the missionaries. The Societas Verbi Divini (SVD), or Society of the Divine Word, started to proselytise along the river in the 1890s (Huber 1988). Evangelists deliberately uprooted local cosmological, ritual and social tenets, and especially targeted the male initiation cult. Today, Tambunum is served by a small cement church located across the river from the main village, adjacent to the primary school. Interestingly, Eastern Iatmul do not practice riverine baptism, despite their use of the Sepik for both everyday and ritual bathing. It is as if the 'natural' watercourse is thought to be unsuitable for the 'civilising' project of Christianity.

The next mention of the Sepik in the annual reports of the New Guinea Company occurs in 1907–08. The river's 'powerful tribes' are reported to have resisted recruiters, missionaries and government agents (Sack and Clark 1979: 277). But Indigenous 'power' was undoubtedly weakened by shipborne diseases such as smallpox and influenza (Crosby 1997). At the same time, there was a growing European presence along the river (Bragge et al. 2006: 102–3). In the early twentieth century, the river was explored by the Südsee Expedition sponsored by the Hamburg Academy of Science, the German–Dutch Border Expedition, the Kaiserin-Augusta-Fluss Expedition and George Dorsey from the Field Museum in Chicago (Reche 1913; Schultze 1914; Behrmann 1922; Schindlbeck 1997; Welsch 2001). In 1909, one German museum director even compared the Sepik to the Friedrichstrasse, a popular shopping district in Berlin (Buschmann 2009: 86). The river and local subjectivities were now thoroughly 'glocal'.

In 1913, the German colonial administration established an outpost at Angoram, which remains one of two administrative centres for the Middle Sepik. The other, Pagwi, is much further upriver. Angoram and Pagwi are the only locations along the entire river with regular public transport by road to Wewak. By the outbreak of World War I, riverine folk were making regular use of European tools, wearing Western garb, speaking rudimentary Tok Pisin and exchanging manufactured porcelain replicas of shell valuables. They greeted every European with art and artefacts in hand (Firth 1982: 163–4; Kaufmann 1985). Yet cross-cultural relations remained fraught, even sometimes brutal (Firth 1982: 96; Schindlbeck 1997: 35; Scaglion 2007: 351). My point in sketching this history is simply to show that, beginning as early as the 1880s, the river facilitated the ongoing flow of disruptive, albeit often celebrated, European novelties, institutions, constraints and privations. So the river is not only the font of local origins, identity and prehistory, but also the source of the encounter with modernity, with all of its promises and frustrations.

During World War I, Australian naval vessels occasionally patrolled the Sepik, but there was little administrative oversight along the river. In consequence, conflicts between local people and German labour recruiters escalated, largely because the latter now felt unrestrained. Some recruiters turned not just physically but also sexually abusive (Hiery 1995: 89). In one instance, a notorious recruiter protested to the administration about an attack on his men by Sepik people—even though he admitting to killing dozens of Melanesians. In response, an Australian military vessel fired on villages up and down the river, while also pilfering artefacts for an Australian museum (ibid.: 88–9). Despite European terror, one Sepik man, having laboured on a plantation in Australia, tried to assert a Queensland identity (ibid.: 101). Local people resisted Europeans—sometimes killing them—even as they drew on European idioms to reconstruct their identity. To 'see' the river today as in some sense pristine, or untouched by the tides of history, is to ignore the currents of a tumultuous past.

After the war, Article 22 of the Treaty of Versailles transferred the colonial administration of New Guinea to Australia as a 'sacred trust', requiring nothing less than the civilising of local people 'not yet able to stand by themselves under the strenuous conditions of the modern world'. In the Sepik, this moral imperative resulted mainly in the continuation of the colonial status quo: labour recruiting, evangelising, amassing art and artefacts, and prospecting for resources. In the 1920s and 1930s, the river was navigated by the Anglo-Persian Oil Company and the Crane Pacific Expedition (Shurcliff 1930; Webb 1997), by orchid hunters and by Beatrice Grimshaw, the unofficial publicist of the Australian colonial administration (McCotter 2006: 85). Several voyagers entered the river simply seeking adventure. They included William Albert Robinson, a young American sailing the world in a 32-foot ketch (Robinson 1932), and Margaret Matches, who popularised her trip along the 'River of Death' in a book called *Savage Paradise* (1931). Even Errol Flynn visited the Sepik, which he described in the *Los Angeles Times* as 'the nearest approach to hell on earth I can conceive'. The unpublished field notes of Gregory Bateson and Margaret Mead (n.d.), who lived in Tambunum for six months in 1938, indicate regular traffic by Europeans with varying commercial and official interests.

World War II proved horribly traumatic in the Sepik (Harrison 2004). In the later 1980s, a slew of older people in Tambunum could still recall gunboat patrols, aerial battles and bombing that destroyed the old men's house, and horrible deprivations under the Japanese military occupation, including summary execution. A few men laboured for the Australian army; many people fled; everybody suffered food shortages. This history is far too complex for a cursory summary. I do note, however, that in 2014, I walked up the village to photograph bomb craters I had seen years before in the bush beside the cult house, but I only found water. The evidence of the war, once so noticeable here, had eroded into the river.

Prior to the arrival of Europeans, Tambunum was enmeshed in regional exchange networks (Gewertz 1983). This trade was multi-ethnic but still localised and bounded. Such insularity, however, ended long ago, despite the common claim that, as Symbiosis Custom Travel (2015) declares, the Sepik remains 'scarcely contacted by the outside world'. As I have shown, the river has facilitated enormous, globalised transformations since the very beginning of the colonial period, allowing for distant influences, people, objects and commodities to flow into the region, and for art, artefacts, labourers and—more recently—timber, cacao, vanilla and 'organic Sepik rice' to flow away. Eastern Iatmul experience the river as a singular force that ceaselessly erodes the 'ground' of society. But they also fault the river for socio-economic sedimentation. Thus, the river simultaneously sustains and subverts modernity.

Cleanliness and Dirt

The Sepik River lacks large-scale commercial enterprises. This absence delights environmentalists and the few tourists who still travel along the river. Travel brochures fulsomely describe the Sepik as mystical and mysterious, 'one of the last unspoiled freshwater ecosystems within Asia … making for a pristine environment not far removed from ancient times' (Remote Lands n.d.). But people in Tambunum voice no such delight in their 'ancient' ways. It is, to be sure, picturesque to observe a group of women cleaning their daily catch in the water (see Figure 8.8), but they would much rather be opening a can or a package.

Figure 8.8 Women cleaning fish.
Source: Photo by the author.

Iatmul, wrote Bateson (1932: 249) almost a century ago, 'are entirely dependent on the great river and the fens for their food and life'. The same is true today. The river provides fish, prawns and mayflies. The river continues to be used for drinking and bathing as well as laundering. Dugout canoes remain the primary means of transportation. But Eastern Iatmul see this continued reliance on the Sepik not as a trophy of noble self-sufficiency and tribal authenticity but as a daily reminder of their thwarted desires to attain modernity, leading them to say that 'we are going backwards'. The 2009–10 floods only served to intensify this despondency by necessitating the rebuilding and replanting of social life just to maintain what they regard as the *same* woeful level of 'development' that they suffered *before* the deluge.

The Sacred Lands Film Project (2017) has announced that the Sepik is the 'soul' of PNG. People in Tambunum voice a more pragmatic assessment of the river: as an untapped resource and, worse, an outright hindrance to development. This is not to say that local people would welcome ecological devastation. They are keenly aware of the perils posed by extractive industries. Nevertheless, Eastern Iatmul bewail the absence of a 'road' to development.

The idiom of a 'road' reflects a traditional view of reality as parsed into multiple paths, each pertaining to different forms of being such as humans, the dead and spirits (see Bateson 1936: 237). It also reflects the many cosmogonic routes of ancestral migration and the criss-crossing of footpaths found in an intact village unharmed by the river. This idiom also speaks to the yearning for a modern road to town and a metaphorical road to modern prosperity (Silverman 2013). Moreover, as one man put it in 2008, Eastern Iatmul must now choose among several, often incompatible, lifestyle 'paths' defined by tradition, the church and things that are 'new' (*kupi* in Iatmul). The ongoing quest for a 'road', however it is variously defined, dethrones the river from its centrality in the local cartography and cosmology. For while Eastern Iatmul proudly affirm their identity as 'Sepiks', they also aspire to 'rise up' to be something more than a mere river people possessing 'only water and fish'.

We can also understand modern ambivalence towards the river with regard to the value of bathing and cleanliness. River folks, as I mentioned earlier, disparage their bush-dwelling neighbours. The Sawos may benefit from timber, but they lack the river, and so remain rank and grubby, say Iatmul, infected with scabies and foolishly content to drink foul water. The Sawos are also inept at swimming and paddling canoes. The river, then, defines Eastern Iatmul as financially wanting but morally and bodily superior to the Sawos. Indeed, Iatmul today see the Sawos as little more than the proverbially unschooled, unworldly *nouveaux riche*.

Major Iatmul rituals typically conclude with bathing. During the famous Naven rite, for example, participants are besmirched with filth, then often make a show of rinsing in the river (Silverman 2001). The Sepik cleanses Iatmul bodies and society. Recently, however, local people have been lamenting their impoverishment through the very same insults that they hurl at their neighbours, saying 'our bodies and clothing are dirty'. This dirt signifies the moral failure to uphold the ideals of hygiene so important to colonial legal regimes and post-colonial notions of civility and citizenship. But this newly observed pollution also scorns the river, once a source of cultural superiority, as now being incapable of cleansing Iatmul persons and polity. And while some Eastern Iatmul advocate relocating the community inland as a solution to riverine erosion, others declare than any such move would irreparably transform Iatmul culture into something 'no better than the Sawos'.

In the late 1980s, tourists regularly disembarked in Tambunum from the *Melanesian Discoverer*, the ship that replaced the *Melanesian Explorer* featured in the film *Cannibal Tours* (Silverman 2013). The boat set sail several times each month from a luxury resort in the coastal town of Madang.[5] At that time, the village received a sizable and regular income from tourists and artefact buyers—especially in comparison to nearby communities. Admittedly, Eastern Iatmul never assessed these revenues as a genuine form of 'development', but the river still brought forth a modicum of modernity. However, these visits ceased when the *Melanesian Discoverer* was sold in 2006. No other tourist boat now regularly plies the middle river. By 2015, another tourist ship that travelled the river, the *Sepik Spirit*, was permanently moored upriver near another Iatmul village (Trans Niugini Tours 2015). It remains in use today. Visitors arrive by charter flight to the Karawari Lodge, located along the Karawari River, but they rarely, if ever, visit Tambunum in smaller 'river truck' boats. In the late 1980s, nearly all men and women in the village were creating some form of tourist art, but such creativity has diminished.

According to local people, the collapse of tourism in Tambunum occurred at the same time as the breakdown in the delivery of 'basic services' by the provincial government. Then came the terrible flood. Eastern Iatmul tend to condense these three misfortunes as manifestations of a single and pervasive decline. Recent statements about communal filth, then, comment on economic marginalisation as much as they do on the river. By judging their bodies, clothing and community as being 'dirty', local people also call into question their status as authentic Iatmul who keep

---

5   See www.mtspng.com/.

clean and bathe in the river. Cleanliness also signals adult competence and care rather than childhood dependence (Silverman 2001: 95, 125–7). Eastern Iatmul fastidiously sweep the plazas by their dwellings, trim the grass, shovel away animal faeces and otherwise remove 'dirt' from public view. Even during the heyday of tourism, local garments were stained and tattered, but villagers rarely remarked on this. Today they often do. By highlighting soiled clothes, as well as the unkempt state of the village and dirt more generally, Eastern Iatmul signal their failure both to attain modernity and to retain their Iatmul identity. These statements also indict the river for failing to sustain the local culture. The Sepik itself is now 'dirty'.

In the aphorism made famous by Douglas (1966), dirt is 'matter out of place'.[6] Recent declarations of social and bodily filth by Tambunum people suggest that the village itself is 'out of place', located neither in the traditional past of unclothed cleanliness, nor in the modern future of laundered clothing. In this liminal space, Tambunum lacks stable and well defined categories. No longer does the once cleansing river provide cultural clarity. Menstruating and post-partum women formerly helped to preserve the 'purity' of the Sepik by bathing in the swamps behind the village, but now they just wash in the river. Many men say that this gendered transgression also contributes to the wider desecration of social life and the natural environment.

Some villagers affix corrugated metal panels to the thatched roofs of their houses to divert rainwater into metal drums or plastic barrels. Unfortunately, these supplies last only a few weeks into the dry season. Everybody therefore relies on the Sepik for drinking water, and most people do so throughout the year. Yet there is widespread recognition that the river harbours parasites and other forms of 'dirt' that are especially deleterious for children. That villagers still drink from the Sepik, and thereby suffer ill health, is another remarked upon sign of their backwardness.

Surprisingly, the Sepik is a 'relatively unproductive fishery', yielding only 10 per cent of the catch from comparable rivers worldwide (Coates 1985, 1993). In fact, the river is 'distinctive because of what is absent, rather than what is present' (Dudgeon and Smith 2006: 207). To remedy this absence, the PNG Government partnered with the United Nations

---

6   Douglas is usually credited with this phrase, but it was also used by Freud and others. While often attributed to Lord Chesterfield, the phrase probably originated with Lord Palmerston (Fardon 2013).

Food and Agricultural Organization on several fish-stocking projects in the 1980s and 1990s, introducing almost 10 exotic species into the Sepik. Previously, other exogenous fish had 'escaped' into the river from harvesting projects elsewhere in the region—most notably tilapia (*Oreochromis mossambicus*) in the 1950s and the common carp (*Cyprinus carpio*) in the 1970s. Although widely consumed by local villagers, these introductions reduced the stock of native species.

Since then, many other exotic fish have colonised the Sepik, such as the Java carp (*Barbonymus gonionotus*) and a species of *Prochilodus* from the Amazon Basin known in Tok Pisin as *raba maus* ('rubber mouth').[7] These fish are now reported to have decimated the fish introduced in earlier decades, which had become staples in the local diet. Several people in 2014 remarked that all the fish I ate in 1988–90 were 'finished'. Now, they said unhappily, 'we only have new fish [or] rubbish fish' that cause intestinal and other ailments. They yearn for the river that once was, saying 'we want the fish we grew up on'. Again, the Sepik is no longer seen as a source of nourishment.

We must also consider the introduction of exogenous plants. In the 1970s, *Salvinia molesta*, a floating fern native to Brazil, was inadvertently introduced into the Sepik, followed in the 1990s by the South American water hyacinth. In both cases, the rapidly proliferating weed choked small watercourses and lakes, obstructed canoes, killed fish and threatened the very livelihood of many communities (Gewertz 1983). In both cases, the Australian Commonwealth Scientific and Industrial Research Organisation (CSIRO) successfully introduced weevils as biological control agents (Thomas and Room 1985; Julien and Orapa 1999). The 'pristine' state of the Sepik, so often celebrated by environmentalists and travellers, rests upon a naïve view of ecological change that mutes the experiences and voices of local people.

Villagers are well aware of the fragility of the Sepik. They also toss the detritus of modernity—like empty food tins, batteries and plastic wrapping—into its waters. In the 1980s, most people casually dismissed any forewarnings about pollution by affirming the power and size of the river to overcome any human activity. Today, Eastern Iatmul voice far less

---

7   For the identification of this fish, I thank Heiko Bleher, via Gerry Allen, who forwarded the identification to Robin Hide, who posted the information on ASAONET@LISTSERV.UIC.EDU on 21 June 2015. See also Dudgeon and Smith (2006: 208).

confidence in its invulnerability, especially with regard to the massive gold and copper mine due to be developed along the Frieda River, an Upper Sepik tributary. The mine, one of the largest in the world, is expected to begin production in 2020, processing 700 million tonnes of material over its initial 17-year lifespan (PanAust 2016). Eastern Iatmul, mindful of similar projects elsewhere in PNG (Kirsch 2006), reasonably fear that the mine will 'pollute' the river. Many villagers already claim that prospecting activities have degraded grasses and reduced some fish species to 'skin and bones'. Several years ago, people in Ambunti, an Upper Sepik administrative centre, similarly attributed a fish kill to chemical pollution from the exploration camp (Colin Filer, personal communication, October 2017).

Additionally, some Eastern Iatmul now tell of plans to cut an enormous channel through the twists and turns of the Sepik. The aim of this project is supposedly to facilitate the efficient passage of barges carrying mineral ore to the sea. I was also told that the 'government' ordered all villages in the floodplain to relocate several kilometres inland to escape the imminent pollution. To add insult to injury, they said, the mining company would only subsidise new housing and food for five years. Neither rumour, of course, is realistic or feasible, yet both stories flag local anxieties over the future capacity of the river to sustain local culture. Both rumours also suggest that the Sepik will fail riverine dwellers while profiting Europeans and elite, non-local Papua New Guineans.

As previously noted, the large volume of water in the river was once seen as an asset that would effectively wash away any possible pollution. Today, ironically, many Eastern Iatmul say that the river was much smaller in the past and has swelled in recent years due to global warming. Similarly, the river is said to flood with greater frequency and devastation. Local people also speak of melting ice caps and rising seas that threaten nearby coastlines (see Lipset 2011), and the unstable, unpredictable timing of the wet and dry seasons. Characteristics of the river that were once beneficial are now seen as liabilities caused by forces that lie well beyond local control or understanding.

Ritual Waters

Recall that the local cosmogony highlights the creation of terrestrial differentiation out of an original aquatic formlessness. River and ground should remain apart—a message conveyed by myths that often situate death and tragedy at the liminal riverbank, where water meets land. But the ornamentation of ceremonial spirits does just the opposite— combining motifs and animals from different topographic realms. In a ceremonial context, these ornaments must shake and blur to recall ocean waves breaking on the shore. Everyday life requires the separation of river and land that otherwise results in dangerous disorder best confined to the spiritual realm.

Figure 8.9 *Aqwi* floating islands.
Source: Photo by the author.

With each rainy season, the surging river regresses the world to the primal sea—albeit in a temporary reversion that regenerates gardens. During these months, the Sepik restricts social life to the interior of dwellings. Houses are likened to clumps of grassy ground or floating islands, called *agwi* in Iatmul, that drift down the river (see Figure 8.9). The king posts of each house are also thought to contain a spirit known as the *agwi* of the house. Houses, too, like most Iatmul artefacts, are decorated with wave motifs (see Figure 8.3). Aesthetically, then, the entirety of human artifice appears to float on the Sepik. In fact, Gregory Bateson (1936: 230) and I, despite the gap of more than half a century separating our respective periods of fieldwork, were both told by Iatmul men that the visible, material world is merely a reflection of water ripples. All efforts to 'ground' social life are therefore ultimately futile, since the true nature of reality is flowing, feminine water.

Mythical and historical narratives often detail the land-based exploits of male ancestors, especially the planting of trees and the construction of villages. Today, the ground is similarly said to be 'built up' by the work of daily life. Human endeavour, in this sense, seeks to stave off riverine erosion. But there is a gendered nuance to the relationship between land and water. The word for 'coconut palm' (*tupma*) also connotes a village. Myth associates coconuts with testicles. Similarly, patrilineages are likened to tall trees, while smaller groups are described as *yarangka* ('branches')—a term that also refers to streams. In general, Iatmul culture encodes a wide-ranging dialogue between the feminine river and terrestrial manhood.

Mortuary ceremonies reveal the same symbolic dialogue (Silverman 2016a). To open the rite, men plant the stalks of totemic plants in front of the cult house. This 'father tree' (*nyait mi*) signifies a male desire for genealogical rootedness in a world threatening to dissolve into the river. But the final scenes of the funeral challenge this symbolism. Mourners wade into the Sepik to wash away their grief. Afterwards, women on the riverbank burn effigies, along with some of the deceased's possessions, and sweep the cinders into the current. The ashes, like the ghosts, are borne out to sea, and to the village of the dead, by ancestral crocodile spirits, floating islands or mystical canoes. The death of both humans and ground entails immersion in the river and, during the funeral, the river itself becomes a liminal path between the living and the dead, and between the land and the sea (see Tuzin 1977). Despite their best efforts to 'build up' the land, the Iatmul always return to the river.

Each evening, Iatmul moor their dugout canoes and modern dinghies at the riverbank. In a similar fashion, men appear to tether the cult house to the 'father-tree' during a funeral. This gesture symbolically prevents the feminine Sepik from washing away the cultural creations of manhood, and yet this gesture ultimately fails to achieve its aim.

There is something graceful about the sight of Eastern Iatmul silently canoeing through the morning mist (see Figure 8.10). Equally elegant is the basso sound of a woman's paddle, intentionally carved at a slight angle, as she dips the blade in the water. Both aquatic images of reverie represent the agile, forward momentum of human agency in the creation and maintenance of social life (see also Krause 2013). The proper role of humanity is to master the river, but nowadays the river masters the community.

Figure 8.10 Man paddling in the morning mist.
Source: Photo by the author.

Some Eastern Iatmul attribute the recent deluges to Mendangumeli, a fierce male crocodile spirit who mythically flooded the world to punish human immorality (Silverman 2016b). All crocodile spirits, regardless of gender, contribute to the masculine ferocity of the river. When Mendangumeli's flood subsided, the spirit slew a young woman. I was recently told that all flood waters seek to kill someone—typically a child—before they recede. Some people allegedly speak to the Sepik directly, asking the river to spare its victim. And yet, as I intimated earlier, the river connotes both death and uterine fertility (Silverman 1997: 113).

Similarly, the ever-flowing and meandering Sepik ceaselessly erodes the terrestrial footing of humanity, but the constant deposition creates sandbars that act as levees sheltering new 'grounds' of social life. Additionally, house posts are only secured in the ground by the annual floods, when the silt acts like mortar. As a fertile source of nurture, the river enables and anchors social life; however, as an aggressive warrior, the Sepik threatens to engulf society in the waters of instability.

Iatmul today cut coffins from old canoes—a modern custom that they attribute to Catholic decorum and colonial sanitary standards. Large canoes represent a key dimension of male personhood, both in terms of the memory of past warfare and in relation to the ongoing practice of chiselling aggressive spirits into the prows. Men proudly affix outboard motors to their canoes; they feel somewhat diminished if they lack a motor and therefore must rely on others to haul cargo or speed along the river. Small canoes are associated with women and therefore with children, food and motherhood (Silverman 2001: 77). All canoes, especially large ones, represent crocodile spirits, and men associate canoes with the reproductive phallus in contrast to uterine water. Crocodile spirits are said to oversee or determine pregnancy. However, as I noted earlier, the downstream journey made by the spirits of the newly dead returns them to the prenatal uterine realm of the ocean (Silverman 2016a). There are numerous ways in which the river, either directly or through its creatures and vehicles, conveys a complex dialogue about the gender of reality.

Conclusion: 'Ball Cutters'

As we have seen, trees, villages and land can take on a masculine identity in contrast to the feminine and uterine associations of the river. However, just as the river is not wholly female, so the land is not wholly male. During one ritual dramatisation of creation, maternal ancestresses, personified as floating islands, lay eggs in the form of land. Some say these mythological isles still support the terrestrial world above the primal sea. Others place the islands on male crocodile spirits, themselves floating on the feminine ocean. But the opposition of the river to the land may be an illusion. Henry Gawi from Tambunum told me that land is female, not male, because men in PNG really only fight over two things—land and women—and from this he concluded that there is no true 'ground', just water, waves and women. And while local women would not necessarily appeal to the first point in Henry's argument, they nonetheless concur with his overall sentiment that femininity, and motherhood in particular, is what sustains social life. That is to say, women in Tambunum interpret masculine bluster as little more than a façade for the pitiful contributions that men make to 'raising up' society.

As part of several ceremonies, and sometimes during the night in the dry season, men perform bamboo flute duets. Women must never glimpse the flautists or the instruments. Men visualise the melodies, said to be the voices of crocodile spirits, as a pair of brothers paddling a canoe. The repetitive, cyclical tunes are named by reference to aquatic imagery, such as brisk currents, swift fish and rivulets of rainwater coursing down from the distant mountains. According to Henry Gawi, the melodies also evoke the hidden truth that reality is a state of watery motion and uncertainty (see Yamada 1997). This is the ultimate truth, if there is one, to Iatmul ontology in Tambunum today.

During the disastrous flood of 2009–10, most people in the community sought refuge in the bush. The deluge destroyed nearly all of their food resources, so that when they returned to the village—or what was left of it—they could only dine on sago and the fish they netted, hooked or trapped. In the 1980s and 1990s, Eastern Iatmul regularly purchased packaged foods, especially rice and tinned mackerel, in village trade stores. At that time, a meal of sago and local fish was seen as traditional but also a sign of indigence. But the decline in revenues from tourism in recent years, coupled with the escalating cost of petrol, resulted in the closure of all the village trade stores. Most Eastern Iatmul now depend again on subsistence gardening, sago harvesting and fishing. They have, as they often say joylessly, gone back to living like their ancestors—and they often attribute this regression to the river. In this sense, the Sepik has reversed the march of time.

Most cash crops, such as coffee and rubber, cannot withstand the annual floods. Vanilla and cacao are exceptions, but financial returns are meagre. Crocodile farms require capital investment in equipment, and so remain largely unfeasible. The recent flood also drowned local confidence in the capacity of local, provincial and national governments—even the wider world—to offer assistance. Indeed, several villagers told me that PNG's prime minister at the time, Michael Somare, stated publicly that 'they are river people, and so are used to this'. Tambunum became a shadow of its former vitality and splendour—not the qualities of some romanticised past of primitive grandeur but of the recent past of only 20 years ago.

The water lines of the recent floods, easily visible on trees and house posts, serve as a constant reminder of the fragility of land-based social life amid a watery fate. Iatmul speak with almost one voice—a rare occurrence under normal circumstances—when they say that they have been defeated by the river and are only left with 'water and fish'.

According to several village men, the source of the 2010 flood was a combination of three things so often said in PNG to bedevil the modern male self: lack of commitment to entrepreneurial spirit, envy of other people's financial successes and inebriated misconduct. According to one story, a hardworking gardener from an upstream village earned a small profit from selling sweet potatoes in town, but a lazy man (*lesman* in Tok Pisin) was envious when the gardener, in typical PNG fashion (their phrase, not mine), spent his earnings on beer and then drunkenly insulted others. The lazy man cast a spell on the river, thus unleashing the flood.

There are mythical sources for this scenario. Long ago, a man was horribly misled by his cross-cousin into killing his wife. Furious at the betrayal, he beseeched Mendangumeli, the fierce crocodile spirit, to exact revenge, and that is why the spirit flooded the world. In one version, the waters gushed from a severed flower said by men to symbolise Mendangumeli's penis. In 2014, when I asked women to comment on this interpretation, they burst into laughter. They also berated men for so obviously yet pitifully trying to assert masculine primacy despite the overwhelming importance of women who give birth to children, work without rest and provide for the community (Silverman 2016b). In private, men agree with them! They confess that women ultimately 'support' and 'sustain' the village. The terrible flood of 2010 similarly attested to the futility of masculine achievement and the destructiveness of male competition. Although men were actively rebuilding the 'ground' of social life after the waters receded, their work was hardly triumphant. There was a sense of dejected resignation to the power of the river.

We can now understand the symbolism of another invasive fish in the Sepik, the pacu (*Piaractus brachypomusor*). This omnivorous fish, originally from the Amazon and closely related to the piranha, is known throughout the Sepik as the 'ball cutter'. The reputation of the pacu has garnered considerable worldwide attention. One example is an April 2011 television episode of Animal Planet entitled 'River Monsters':

Jeremy Wade [host of the show] travels to Papua New Guinea to investigate a spate of bizarre deaths on one of the world's last great unexplored rivers, the Sepik. A creature there is tearing chunks from unsuspecting fishermen and devouring certain male body parts. Jeremy has never fished in this part of the world and knows little of what might be there. Nothing could prepare him for what he discovers. (Animal Planet 2011)

This 'testicle-eating fish' now purportedly menaces bathers from Scandinavia to Texas.

The pacu was introduced into the Sepik in the 1990s as yet another potential food source, but it rapidly denuded the local flora with the massive teeth that it uses to crunch berries and seeds, and contributed to a further decline of native fish species (Bell et al. 2011: 587). Eastern Iatmul report that the 'ball cutter' specifically consumes the eggs and hatchlings of these indigenous species. Scientists are sceptical about the alleged testicular attacks (Dau 2001), but that is beside the point. This is just another example of the multiple ways in which the river has betrayed the people of Tambunum. The very name 'ball cutter' certainly speaks to men's anxieties today at the sense of being emasculated by the river.

In the coffee-table book *Cousteau's Papua New Guinea Journey*, the chapter on 'River of the Crocodile Men' describes the Sepik in redolent prose as a 'fluid prairie' and a 'rubbery world' of 'uncommon quiet' where '[l]ife changes ... and life does not change at all' (Cousteau and Richards 1989: 131). But the Cousteau expedition phrased their account as an affirmation of an earlier, simpler era of spirits, ritual and myth. In Tambunum today, such a contradictory statement is less like a celebration of an enduring cultural tradition and more like a resigned acknowledgment of the failure of social and economic progress, akin to paddling a canoe against the current and getting nowhere. Like river water, social reality in the Middle Sepik today offers nothing stable to grasp.

Acknowledgments

Over the years, many institutions have graciously supported my fieldwork: the Fulbright Foundation, the Institute for Intercultural Studies, and the Graduate School and Department of Anthropology at the University of Minnesota, all in 1988–90; the Wenner-Gren Foundation for Anthropological Research and DePauw University in 1994; Wheelock College in 2008; Wheelock College and the Institute for Money,

Technology and Financial Inclusion at the University of California at Irvine in 2010; and Wheelock College, through a Gordon Marshall Fellowship, in 2014. I also acknowledge the Women's Studies Research Center at Brandeis University, not only for my affiliation but also for reminding me of the importance of gender in all my work. Needless to say, my gratitude to the people of Tambunum remains profound and inevitably unrequited, especially during their current plight. For this, I apologise. I also acknowledge the helpful comments and collegiality of participants at the conference session from which this volume is derived, especially our organiser, John Wagner, and his co-editor, Jerry Jacka.

References

Animal Planet, 2011. 'River Monsters.' Viewed 16 February 2017 at: www.animal planet.com/tv-shows/river-monsters/

Bateson, G., 1932. 'Social Structure of the Iatmul People of Sepik River.' *Oceania* 2: 245–291, 401–453. doi.org/10.1002/j.1834-4461.1932.tb00029.x

——, 1936. *Naven: A Survey of the Problems Suggested by a Composite Picture of the Culture of a New Guinea Tribe Drawn from Three Points of View*. Cambridge: Cambridge University Press.

Bateson, G. and M. Mead, n.d. Fieldnotes, Tambunum Village, Sepik River, New Guinea, 1938. Washington (DC): Library of Congress, Margaret Mead Papers.

Behrmann, W., 1922. *Im Stromgebiet des Sepik: Eine Deutsche Forschungsreise in Neuguinea*. Berlin: August Scherl.

Bell, J.D., J.E. Johnson and A.J. Hobday, 2011. *Vulnerability of Tropical Pacific Fisheries and Aquaculture to Climate Change*. Nouméa: Secretariat of the Pacific Community.

Bragge, L., U. Claas and P. Roscoe, 2006. 'On the Edge of Empire: Military Brokers in the Sepik "Tribal Zone".' *American Anthropologist* 33: 100–113. doi.org/10.1525/ae.2006.33.1.100

Buschmann, R.F., 2009. *Anthropology's Global Histories: The Ethnographic Frontier in German New Guinea, 1870–1935*. Honolulu: University of Hawai'i Press.

Claas, U., 2009. '"Fish, Water, and Mosquitoes": The Western Invention of Iatmul Culture.' In E. Hermann, K. Klenke and M. Dickhard (eds), *Form, Macht, Differenz: Motive und Felder: Ethnologischen Forschens.* Göttingen: Universitätsverlag Göttingen.

Claas, U. and P. Roscoe, 2009. 'Manuscript XXI: A Journey up the Sepik River in 1887.' *Journal of Pacific History* 44: 333–343. doi.org/10.1080/00223340903358431

Coates, D., 1985. 'Fish Yield Estimates for the Sepik River, Papua New Guinea, a Large Floodplain System East of "Wallace's Line".' *Journal of Fish Biology* 27: 431–443. doi.org/10.1111/j.1095-8649.1985.tb03191.x

——, 1993. 'Fish Ecology and Management of the Sepik-Ramu, New Guinea, a Large Contemporary Tropical River Basin.' *Environmental Biology of Fishes* 38: 345–368. doi.org/10.1007/BF00007528

Cousteau, J.-M. and M. Richards, 1989. *Cousteau's Papua New Guinea Journey.* New York: Harry N. Abrams.

Crosby, A.W., 1997. 'Papua New Guinea, its Demographic History and Infectious Diseases.' In H. Hiery and J.M. Mackenzie (eds), *European Impact and Pacific Influence: British and German Colonial Policy in the Pacific and the Indigenous Response.* London: I.B. Tauris.

Dau, J., 2001. 'Sepik and Ramu Rivers "Killer Fish" Not Pacu, Says Aquaculturist Middleton.' *The National*, 20 June.

Douglas, M., 1966. *Purity and Danger: An Analysis of Concepts of Pollution and Taboo.* London: Routledge & Kegan Paul. doi.org/10.4324/9780203361832

Dudgeon, D. and R.E.W. Smith, 2006. 'Exotic Species, Fisheries, and Conservation of Freshwater Biodiversity in Tropical Asia: The Case of the Sepik River, Papua New Guinea.' *Aquatic Conservation* 16: 203–215. doi.org/10.1002/aqc.713

Fardon, R., 2013. 'Citations out of Place.' *Anthropology Today* 29(1): 25–27. doi.org/10.1111/1467-8322.12007

Feldhaus, A., 1995. *Water and Womanhood: Religions Meanings of Rivers in Maharashtra.* New York: Oxford University Press.

Firth, S., 1982. *New Guinea under the Germans.* Melbourne: Melbourne University Press.

Fortune, K., 1998. *Malaguna Road: The Papua New Guinea Diaries of Sarah Chinnery.* Canberra: National Library of Australia.

Full, A., 1909. 'Eine Fahrt auf dem Kaiserin Augustafluß.' *Deutsches Kolonialblatt: Amtsblatt für die Schutzgebiete in Afrika und in der Südsee* 20: 739–741, 744–745.

Gewertz, D.B., 1983. *Sepik River Societies: A Historical Ethnography of the Chambri and their Neighbors.* New Haven (CT): Yale University Press.

Harrison, S., 2004. 'Forgetful and Memorious Landscapes.' *Social Anthropology* 12: 135–151. doi.org/10.1017/S0964028204000436

Hiery, H.J., 1995. *The Neglected War: The German South Pacific and the Influence of World War I.* Honolulu: University of Hawai'i Press.

Huber, M.T., 1988. *The Bishops' Progress: A Historical Ethnography of Catholic Missionary Experience on the Sepik Frontier.* Washington (DC): Smithsonian Institution Press.

Julien, M.H. and W. Orapa, 1999. 'Structure and Management of a Successful Biological Control Project for Water Hyacinth.' In M.P. Hill, M.H. Julien and T.D. Center (eds), *Proceedings of the 1st IOBC Global Working Group Meeting for the Biological and Integrated Control of Water Hyacinth, 16–19 November 1998, Harare, Zimbabwe.* Pretoria: Plant Protection Research Institute.

Kaufmann, C., 1985. 'Postscript: The Relationship between Sepik Art and Ethnology.' In S. Greub (ed.), *Authority and Ornament: Art of the Sepik River, Papua New Guinea.* Basel: Tribal Art Centre.

Kirsch, S., 2006. *Reverse Anthropology: Indigenous Analysis of Social and Environmental Relations in New Guinea.* Stanford (CA): Stanford University Press.

Krause, F., 2013. 'Rapids on the "Stream of Life": The Significance of Water Movement on the Kemi River.' *Worldviews* 17: 174–185. doi.org/10.1163/15685357-01702008

Leigh, C. and R. Perry, 2011. *Art Dealer in the Last Unknown: Ron Perry and New Guinea Art, the Early Years, 1964–1973.* Tucson (AZ): Carolyn Leigh Studios.

Lipset, D.M., 2011. 'The Tides: Masculinity and Climate Change in Coastal Papua New Guinea.' *Journal of the Royal Anthropological Institute* (NS) 17: 20–43. doi.org/10.1111/j.1467-9655.2010.01667.x

Lipset, D.M. and E.K. Silverman, 2005. 'Dialogics of the Body: The Moral and the Grotesque in Two Sepik River Societies.' *Journal of Ritual Studies* 19: 17–52.

Lutkehaus, N.C., 1995. *Zaria's Fire: Engendered Moments in Manam Ethnography*. Durham (NC): Carolina Academic Press.

Matches, M., 1931. *Savage Paradise*. New York: Century.

McCotter, C., 2006. 'Seduction and Resistance, Baptism and "Glassy Metaphorics": Beatrice Grimshaw's Journeys on Papua's Great Rivers.' *Hecate* 32: 81–97.

PanAust, 2016. 'Frieda River Copper-Gold Project Feasibility Study Completed.' Viewed 23 February 2018 at: www.panaust.com.au/sites/default/files/FRP CompanyAnnouncement%2019%20May%202016_1.pdf

Reche, O., 1913. *Der Kaiserin-Augusta-Fluss*. Hamburg: L. Friederichsen.

Remote Lands, n.d. 'Preferred Hotels: Papua New Guinea—Sepik Region.' Viewed 17 February 2017 at: www.remotelands.com/preferhotel/papua-new-guinea_129_Sepik-Region

Robinson, W.A., 1932. *10,000 Leagues over the Sea*. New York: Harcourt, Brace.

Sack, P. and D. Clark (eds), 1979. *German New Guinea: The Annual Reports*. Canberra: Australian National University Press.

Sacred Lands Film Project. 2017. 'Sepik River Basin.' Viewed 1 June 2018 at: sacredland.org/sepik-river-basin-papua-new-guinea/

Scaglion, R., 2007. 'Multiple Voices, Multiple Truths: Labour Recruitment in the Sepik Foothills of German New Guinea.' *Journal of Pacific History* 42: 345–360. doi.org/10.1080/00223340701692064

Schindlbeck, M., 1997. 'The Art of the Head-Hunters: Collecting Activity and Recruitment in New Guinea at the Beginning of the Twentieth Century.' In H. Hiery and J.M. Mackenzie (eds), *European Impact and Pacific Influence: British and German Colonial Policy in the Pacific and the Indigenous Response*. London: I.B. Tauris.

Schultze, L., 1914. *Forschungen im Innern der Insel Neuginea: Bericht des Führers über die Wissenschaftlichen Ergebnisse der Deutschen Grenzexpedition in das Westliche Kaiser-Wilhelmsland 1910*. Berlin: E.S. Mittler.

Shurcliff, S.N., 1930. *Jungle Islands: The 'Illyria' in the South Seas*. New York: Putnam.

Silverman, E.K., 1997. 'Politics, Gender, and Time in Melanesia and Aboriginal Australia.' *Ethnology* 36: 101–121. doi.org/10.2307/3774078

——, 2001. *Masculinity, Motherhood, and Mockery: Psychoanalyzing Culture and the Iatmul Naven Rite in New Guinea*. Ann Arbor: University of Michigan Press.

——, 2003. 'High Art as Tourist Art, Tourist Art as High Art: Comparing the New Guinea Sculpture Garden at Stanford University and Sepik River Tourist Art.' *International Journal of Anthropology* 18: 219–230. doi.org/10.1007/BF02447907

——, 2013. 'After Cannibal Tours: Cargoism and Marginality in a Post-Touristic Sepik River Society.' *Contemporary Pacific* 25: 221–257. doi.org/10.1353/cp.2013.0031

——, 2016a. 'Funerary Failures: Traditional Uncertainties and Modern Families in the Sepik River.' In D. Lipset and E.K. Silverman (eds), *Mortuary Dialogues: Death Ritual and the Reproduction of Moral Community in Pacific Modernities*. New York: Berghahn Books.

——, 2016b. 'The Waters of Mendangumeli: A Masculine Psychoanalytic Interpretation of a New Guinea Flood Myth—and Women's Laughter.' *Journal of American Folklore* 129: 171–202. doi.org/10.5406/jamerfolk.129.512.0171

Strang, V., 2006. 'Fluidscapes: Water, Identity and the Senses.' *Worldviews* 10: 147–154. doi.org/10.1163/156853506777965802

Swadling, P. and R. Hide, 2005. 'Changing Landscape and Social Interaction: Looking at Agricultural History from a Sepik-Ramu Perspective.' In A. Pawley, R. Attenborough, R. Hide and J. Golson (eds), *Papuan Pasts: Investigations into the Cultural, Linguistic and Biological History of the Papuan Speaking Peoples*. Canberra: The Australian National University, Research School of Pacific and Asian Studies (Pacific Linguistics 572).

Symbiosis Custom Travel, 2015. 'Exploring the Sepik River.' Viewed 16 February 2017 at: www.symbiosis-travel.com/papua-new-guinea/anthropology-tribes/293/exploring-the-sepik-river/

Thomas, P.A. and P.M. Room, 1985. 'Toward Biological Control of Salvinia in Papua New Guinea.' In E.S. Delfosse (ed.), *Proceedings: 6th International Symposium on Biological Control of Weeds, August 19–25, 1984*. Vancouver: Agriculture Canada.

Trans Niugini Tours, 2015. 'Sepik Spirit.' Viewed 6 June 2018 at: www.pngtours.com/lodge4.html

Tuzin, D.F., 1977. 'Reflections of Being in Arapesh Water Symbolism.' *Ethos* 5: 195–223. doi.org/10.1525/eth.1977.5.2.02a00060

UNDP (United Nations Development Programme), 2012. 'Sepik Wetlands Management Initiative.' Viewed 16 February 2017 at: www.pg.undp.org/content/papua_new_guinea/en/home/ourwork/environmentandenergy/successstories/sepik-wetlands-management-initiative.html

Webb, V.-L., 1997. 'Official/Unofficial Images: Photographs from the Crane Pacific Expedition, 1928–1929.' *Pacific Studies* 20(4): 103–124.

Welsch, R.L., 2001. 'One Time, One Place, Three Collections: Colonial Processes and the Shaping of Some Museum Collections from German New Guinea.' In M. O'Hanlon and R. Welsch (eds), *Hunting the Gatherers: Ethnographic Collectors, Agents, and Agency in Melanesia 1870s–1930s*. New York: Berghahn Books. doi.org/10.2307/j.ctt1x76fh4.12

Wikipedia, 2015a. 'List of Rivers by Length.' Viewed 16 February 2017 at: en.wikipedia.org/wiki/List_of_rivers_by_length

——, 2015b. 'List of Largest Unfragmented Rivers.' Viewed 16 February 2017 at: en.wikipedia.org/wiki/List_of_largest_unfragmented_rivers

Yamada, Y., 1997. *Songs of Spirits: An Ethnography of Sounds in a Papua New Guinea Society.* Boroko: Institute of Papua New Guinea Studies.

9. Rivers of Memory and Forgetting

JOHN R. WAGNER

Introduction

Kamu Yali[1] is a water world. Average rainfall is about 5,000 mm per year and daily showers are common even during the dry season that lasts from November to April. The village is located along the shoreline of Nasau Bay in Papua New Guinea (PNG) and its territory extends as far out into the Solomon Sea as village men can safely travel in outrigger canoes. Village houses are built facing the sea, several feet off the ground, in order to avoid wet season floods; local transportation systems are water-based since there are no roads in this part of PNG; outrigger canoes and motorised boats line the beach in front of the houses; villagers go to sleep listening to the pounding of the surf. Inland, behind the village to the south-west, lies a mountain ridge known in the local Kala language as Batatalã ('Blue Mountain').[2] Batatalã is home to a craggy, inhospitable moss-cloud

---

1   The official name of the village is Lababia but in 2002 villagers requested that I refer to it as Kamu Yali in my publications. In their language (Kala), Kamu Yali is the name of one of the hamlets near the centre of the current village site, and is considered by them to be the correct name for the village. They first began using this name in 1996 when they partnered with Village Development Trust (VDT), a local non-governmental organisation, to create a conservation and development project. VDT staff incorrectly wrote the name as Kamiali, however, and this name is still in use today by many people, especially those who first came to know the village through the Kamiali Conservation and Development Project.

2   The spelling of Batatalã and of other local names and words is based on the Kala orthography developed by Schreyer and Wagner (2013) in partnership with the Kala Language Committee. The diacritic that appears above the final letter 'a' indicates a nasal vowel. The letter 'l' is used to indicate an alveolar lateral flap.

forest that blocks the movement of people between high-elevation inland valleys and coastal regions; the headwaters of the four rivers that dominate Kamu Yali territory all arise on the eastern, seaward side of this ridge (see Figure 9.1).

Figure 9.1 Village shoreline dotted with canoes, with Blue Mountain foothills shrouded in clouds.
Source: Photo by the author.

Although Kamu Yali residents, like all Kala people, identify themselves as coastal people (*pipol bilong solwara* in Tok Pisin[3]) and are careful to distinguish themselves from 'bush' or inland people (*pipol bilong bus*), I focus in this chapter on the importance of rivers to Kamu Yali social life and cultural identity. Villagers' reliance on rivers is sufficient for them to be described as a river people, though this is not a claim I have ever heard them make, not even those of them who locate their gardens several kilometres inland from the sea along the Bitoi River. Only a very few villagers are familiar with the upper reaches of any of the rivers, and inland travel beyond gardening areas is infrequent and mainly limited to

---

3   Tok Pisin, generally referred to as Melanesian Pidgin in English, serves as a lingua franca throughout the country. Having become a relatively fluent speaker during the full year I spent in Kamu Yali in 1998–99, it is the language I normally use when in the village.

hunting and the gathering of wild foods, fuelwood and building materials. My focus on Kamu Yali rivers allows me, nevertheless, to demonstrate that coastal people can simultaneously be river people, and that the opposition between salt water and fresh water, though deeply embedded in the cosmology and discourse of many Pacific island peoples (see Hviding, this volume), is historically and geographically contingent rather than universal. It is the combination of agricultural land with riparian and marine resources that makes life viable in this part of the country, and Kamu Yali history and identity are rooted equally in these three domains.

My foregrounding of rivers and riverine resources also allows me to emphasise the agency of rivers within the social lives of Kamu Yali villagers. As noted in the introduction to this volume, scholars from many disciplinary backgrounds now attribute agency to a wide variety of non-human 'actors', a trend that stands in sharp opposition to the more common historical tendency to define agency as an exclusively human characteristic dependent on conscious intent. Social scientists in the field of science and technology studies have been especially prominent among those who apply the term to a wide range of 'actants' (Latour 2005), but they are joined by many others, including ethnographers (Viveiros de Castro 1998; Descola 2013), philosophers (Barad 2003; Bennett 2010; Morton 2013) and archaeologists (Hodder 2012; Malafouris and Renfrew 2013; Edgeworth 2014) who seek to challenge existing ideas about materiality itself as well as the relationship of ecology to culture. It is not my intent here to endorse any particular line of argument on this issue or to present empirical evidence in support of the idea that rivers do, in fact, possess agency. I use the term simply because it helps bridge the gap between social and ecological domains.

> There can, then, be no radical break between social and ecological relations; rather, the former constitute a subset of the latter. What this suggests is the possibility of a new kind of ecological anthropology, one that would take as its starting point the active, perceptual engagement of human beings with the constituents of their world—for it is only from a position of such engagement that they can launch their imaginative speculations concerning what the world is like. (Ingold 2000: 60)

The approach Ingold proposes is also one that constructs an understanding of things based on 'their position within a relational field' rather than on the mere unfolding of internal characteristics (ibid.: 97). Ingram (2010) proposes a relational definition of agency on the basis of his study of human/microbial relations. Following Barard (2003), he proposes that we 'should move away from a focus on the non-human as a separate, ethically and agentially equivalent subject and … focus on the activity that results through relationships and interaction' (Ingram 2010: 102). This is the approach I follow in this chapter as I describe the relationship of Kamu Yali villagers to their four rivers.

More fundamentally, framing my account of Kamu Yali around the rivers also allows me to analyse and interpret historical and contemporary patterns of change as the consequence of selective processes of memory and forgetting. As Harrison (2004) points out in his account of the Manambu village of Avatip, located on the Sepik River in PNG, river features can act as mnemonic devices, playing important roles in the way individual and collective memories are organised and retained. In Kamu Yali, each river is strongly associated with a particular kinship group in the village and is the site of historical events that have shaped the history of the community in distinct ways. Stories about the rivers are thus simultaneously stories about pivotal moments in village history and about the coming into being of the kin groups that continue to structure village life in fundamental ways. But Harrison (2004: 149) also notes that both remembering and forgetting can occur as expressions of power, not just as innocent or random occurrences. This insight also has traction at Kamu Yali, where the mouths of the Alewili and Saia rivers are the sites of proposed chromium mines, and villagers have been asked to identify the respective 'landowners' as part of the approval process for obtaining a mining licence. This process requires that villagers remember and re-engage with ancestral knowledge about the site, but, at the same time, pressures them to renounce and forget that knowledge so that the mining company can transform a place inhabited by ancestor spirits into a zone of industrial exploitation.

The Rivers

Although I was intensely conscious of living in a water world during the period of my doctoral research in Kamu Yali in 1998–99, rivers were by no means the primary focus of my research. I had come to Kamu Yali to study community-based resource management practices, and my main focus was on land and sea tenure, gardening and fishing practices, and the sustainable development projects that were being implemented in the village as part of the Kamiali Conservation and Development Project

(Wagner 2002, 2005). The other most prominent 'resource' on which I concentrated was Batatalã ('Blue Mountain'). Village Development Trust (VDT), the non-governmental organisation managing the project, had begun to emphasise the unusual ecological features of this high-elevation forest in their funding applications, and had facilitated its inclusion in a newly established wildlife management area (Bein 1998; Wagner 1999).

I nevertheless learned a great deal about Kamu Yali rivers though my study of land tenure, since rivers sometimes served as boundary markers, and through my study of agriculture, since the ecological characteristics of rivers largely determined the location of swidden gardens. I also noted many of the other ecological services they provided, such as habitat for wild food species, most notably sago palm. More by chance than by design, I recorded a number of stories about the relationships of rivers to various kin groups in the village. I came to know Kamu Yali rivers most intimately, however, through my own direct and unavoidable physical experiences of them, since foot travel throughout the village territory requires their constant fording. I had to develop strategies that allowed me to cross rivers safely while carrying notebooks, a camera, survey and GPS equipment and, perhaps, a change of clothing depending on the weather and time of year. Travel to distant gardens, or to forested or swampy areas, also required sturdy boots that had to be laboriously unlaced and re-laced for each crossing. The ideal trousers and shirts were ones made from materials that would dry quickly. During the dry season, when the rivers were no more than waist high and only 10 m across, the risks were low, the inconvenience was minimal and they were compensated by the pleasure of a cool swim on a hot day. There was no compensation during wet season, however, which lasted from May until September. Currents could be very strong and the water could be shoulder high or over our heads in the deepest channels. Under those conditions, we had to cross the rivers diagonally, moving downstream with the flow of water, slowly edging towards the other bank while holding our equipment above our heads. The meandering courses of the rivers also meant that we often had to cross the same river several times during a single trip if we were travelling inland away from the mouth of a river. Crossings that were easy for my village companions were always much more difficult for me, and although I had very few spills in the course of a year, these crossings were among my least favourite village experiences.

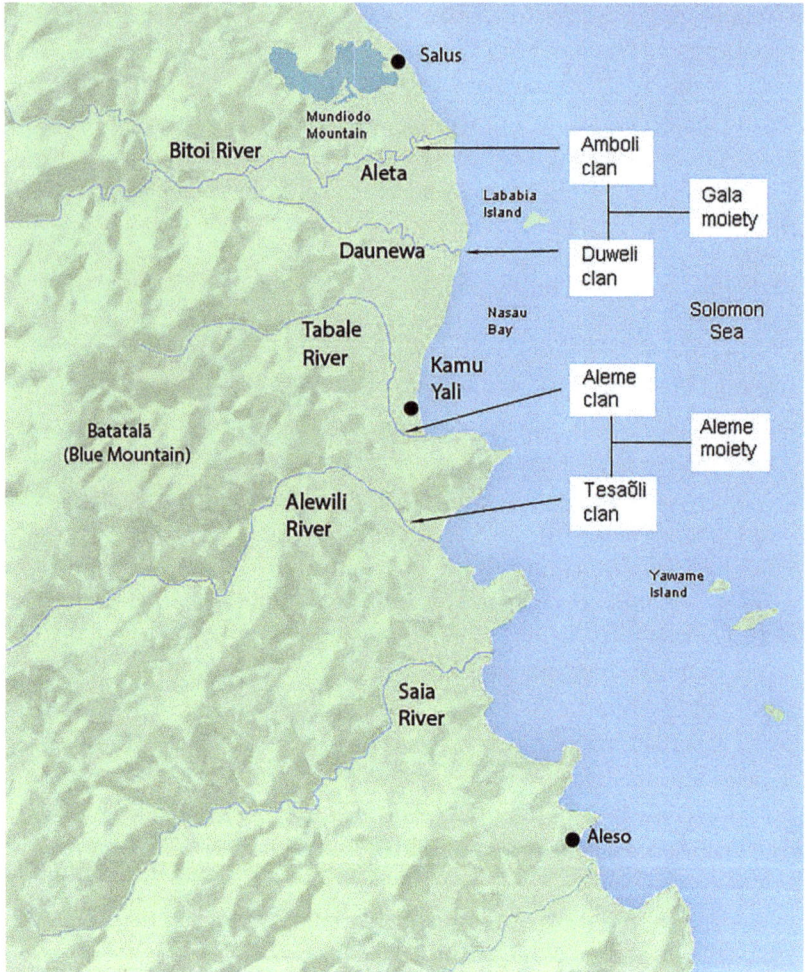

Figure 9.2 Kamu Yali rivers and associated kin groups.
Source: Map by the author.

Of the four main rivers that drain the steep flanks of Blue Mountain, the Bitoi River[4] in the north carries the largest volume of water. As it descends from headwaters that arise at elevations of over 2,000 m, it is joined by

4   The origin of the term Bitoi is uncertain, but dates back to the early 1900s when German colonial officials began assigning names to village sites and local geographical features. Maps of the area still display German names for some features, such as Nasau Bay and Hessen Bay, while other features have names that are neither German nor Kala in origin. A few knowledgeable villagers suggested to me that the official name for the village, Lababia, arose as the result of a misunderstanding by German officials of the local Kala name for the village site they occupied at the time of contact. There is no consensus, however, over the origins of either of the names Lababia or Bitoi.

other streams and by the equally large Buyawim River at an elevation of about 600 m (see Figure 9.2). Shortly after its confluence with the Buyawim, it divides into two distinct channels at the western, inland edge of a broad flood plain that provides villagers with their largest expanse of agricultural land. Villagers refer to the northern and southern branches as Bitoi One and Bitoi Two respectively when speaking Tok Pisin, but in Kala the northern channel is known as Aleta and the southern channel as Daunewa. Both branches have changed their course significantly over the past few decades, and although Aleta now carries a significantly larger volume of water than Daunewa, this was not always the case.

The Tabale River descends behind the village, near the middle of Kamu Yali territory, and meanders through an extensive sago swamp before entering Nasau Bay. The Alewili River is located 10 km further south, well away from the main village and the main gardening areas, and is used mainly as a fishing camp. It descends much more abruptly into the sea than do the northern rivers, over steeper and rockier ground with less agricultural potential. Chromite deposits have been discovered at the mouth of the Alewili, and some families, in anticipation of future mining activities, have begun to plant crops and occupy the site more regularly in order to assert their claims as primary landowners in the area. Chromite deposits have also been discovered at the mouth of the Saia River, which lies a further 10 km south and serves as an informal boundary between the territories of Kamu Yali and Buso, another Kala village. In the account that follows, I describe the four main rivers of Kamu Yali in the order that I came to know them, beginning with the Tabale.

The Tabale River

My decision to conduct doctoral research in Kamu Yali came about as the result of information provided to me by VDT and the permission granted to me by the Kamiali Conservation Project Committee (KCPC), a village committee created to work with VDT on sustainable development projects in the village. One project involved the construction of a training centre and village guesthouse, and VDT arranged for me to stay in the guesthouse during my year in the village. The guesthouse is located in a secluded spot at the southern end of the main village, and is separated from the village proper by the Tabale River. As long as I was willing to get wet, I could easily wade across the river mouth where sand bars were constantly forming and reforming and the water was rarely more than waist high. Otherwise, if I wished to stay dry, I could approach one of the houses beside the river and ask to be ferried across in an outrigger canoe (see Figure 9.3). This was not always convenient, however, and local dogs could become quite aggressive if I approached an unoccupied home on my own. I was also cautioned never to wade across the river at night when crocodiles would occasionally frequent the river mouth. The Tabale River thus restricted my ability to spend time in the village observing daily activities and getting to know people other than those who came to visit me at the guesthouse.

Figure 9.3 Ferry canoe at the mouth of the Tabale River.
Source: Photo by the author.

The Tabale has a very different character at different locations along its course. Several kilometres upstream from the village, at higher elevations, the riverbed is rocky and the water is clear and cool, but after completing its abrupt descent from Batatalā, it reaches a lowland shelf fronting Nasau Bay and meanders for 2 km through a dense mud-bottomed swamp. By the time it reaches the sea, most of the sediment has settled out, and the water is mostly clear where it courses over the sandbars at the river's mouth. At one point in its meanderings, it passes very close to the northern end of the village and, despite the murky quality of the water, this is where many villagers formerly came to bathe, wash their dishes and clothes, and fetch water for home use. These practices came to an end during my first few weeks in the village in 1998 when a new piped water system was completed with the assistance of VDT. The new gravity-fed water system brings water to a series of standpipes installed throughout the village, one for about every eight houses. Village women especially appreciate the convenience of the standpipes since they formerly had to make at least two trips to the river each day to fetch water, an especially onerous chore for those who live a kilometre or more away from the river. This piped water does not come from the Tabale River itself but from a smaller, cleaner tributary that joins the Tabale near its mouth.

The current village site extends along the shoreline for a distance of nearly 2 km. Originally it was composed of a series of separate hamlets, each occupied by a distinct kin group, but most hamlets are now joined to form one continuous village, and the residential boundaries of kinship groups have grown less distinct. The upper or northern end of the village is still referred to as Bik Ples ('Big Place' in Tok Pisin) because of its denser population and its historic status as the main village site. Bik Ples is still occupied predominantly by members of the Gala moiety; before the piped water system was constructed, the Gala used the northern meander of the river as their main source of fresh water. Most members of the other moiety, the Aleme, who were scattered among several smaller hamlets lying south of Bik Ples, would go to a spot at the southern end of the village near the mouth of the Tabale River. The original water use pattern thus emphasised the separateness of the two moieties and provided a daily opportunity for reaffirmation of that separateness through the gatherings at two distinct sites at the beginning and end of each day. The piped water system has thus become one of several factors contributing to an ongoing reconfiguration of kin group identities and social networks.[5] Moieties perform far fewer functions in the village today than in the recent past and are cross-cut by many new institutional forms, such as school, church and development committees.

---

5   See Hviding (this volume) on the cultural significance of a piped water system in the Solomon Islands.

Despite ongoing changes to kin group identities, the Tabale River continues to hold deep historical significance for the Aleme moiety and for one of its most senior clans,[6] which is also known by the name Aleme. Alternatively, both the moiety and its senior clan are referred to as Tabaleli, emphasising their historical relationship to the river. The Kala suffix *li* is a plural marker that is added to kinship terms such as Tabale in order to designate the entire group or clan. The name Aleme appears to be more recent than Tabaleli, for both the senior clan and the moiety, and was given to them by the Gala. A well-known story told by both Gala and Tabaleli individuals concerns the historical relationship of the two groups during a formative period in their history. Gala history is linked to the Bitoi River, and their original village sites were located either beside one of the two arms of the river or on a small island (Lababia) near the mouth of the southern channel of the river. The ancestors of the Tabaleli, on the other hand, lived near the mouth of the Tabale River. Whereas the Gala were primarily growers of taro, the Tabaleli were mainly dependent on sago and fish. The Gala jokingly referred to the Tabaleli as 'Aleme Aleme', because the Kala word *me* means 'and' while the word *ale* means 'water', so the literal meaning of the name is 'and water and water'. The name Aleme thus came to take on the meaning of 'people who go always to water', whether to fish along the fringing reef or to gather sago from the swamp alongside the river.

---

6   I use the English term 'clan' to refer to the named kinship groups that collectively make up each moiety in deference to the practice of Kamu Yali villagers when speaking or writing in English. English usage follows Tok Pisin usage in which *klan* ('clans') are distinguished from *meja klan* ('major clans'). I use the term 'moiety' rather than 'major clan' because, in anthropological terminology, the two larger groups function as moieties but do not constitute unilineal descent groups. The groups villagers refer to as 'clans' are best understood as cognatic rather than unilineal descent groups, and for that reason I have referred to them as 'ramages' in a previous publication (Wagner 2007: 30). Kala villagers use the suffix *li* to designate all kin groups—whether moieties, clans or smaller groups. There are no simple terms to distinguish groups of different size or status from one another; such distinctions can be made in Kala but require lengthy explanations. Kamu Yali thus represents an example of what McKellin (1991) has referred to as 'the pidginization of kinship terminology', reflecting a series of changes that have occurred since the colonial era and that now obscure the nature of pre-colonial kinship practices. The cognatic structure of Kamu Yali helps to explain another important feature of their kinship system, which is the ability to claim multiple clan affiliations and to switch one's primary affiliation if one so chooses. Burton (1996) has described a similar pattern among the neighbouring Biangai. Some Kamu Yali parents assign each of their children to one of their own parents' clans, so that a group of siblings can be distributed among as many as four different clans.

The Bitoi River

Since my primary research goal in Kamu Yali in 1998–99 was to understand the ways in which villagers managed key resources, such as agricultural land, fisheries and forests, I began by trying to understand the rules of the land tenure system as it applied to agricultural land. The research proposal I sent to the village before my arrival included a request to map the distribution of agricultural land by landowning group, which was approved by a village committee. It did not take long to find out, however, that land boundaries were highly contested and that competing groups within the village were interested in supporting their claims through my mapping project. I therefore dispensed with the idea of mapping the lands of named groups in favour of a household-level study. Over a period of a few months, I was able to visit the swidden gardens of 26 households, interview male and female heads of household about their history of use of specific garden plots, survey the size of each of their gardens, and plot garden locations using GPS equipment. Families explained to me in Tok Pisin that each plot of land had a *papa* or *mama bilong graun*, who is the person considered to be the custodian of the ground by virtue of his or her senior position within an extended family or larger landowning group. This research approach revealed that most garden land was under the immediate control of relatively small, extended family groups that had flexible and somewhat unpredictable relationships with the named kinship groups in the village. Everyone agreed that the named groups held underlying rights to specific garden areas, but many areas were subject to competing claims, and some areas were claimed to be *pablik graun* ('public ground') for all to use. Most confusing for me, in the initial stages of research, was the lack of a clear relationship between the clan affiliation of the household head and the clan affiliation of the *papa* or *mama bilong graun* of each garden plot the family was using. Families had access to land through both matrilineal and patrilineal lines of succession and through all four grandparents of the male and female heads of household. Access to land was determined less by the type of kinship tie than by the personal relationship of the person using the land to the custodian of the land—a relationship that had to be constructed through the regular sharing of labour and other forms of gift giving and mutual support.

Despite the fluidity and complexity of the land tenure system, one fact became readily apparent to me during the first few weeks of my agricultural study. The vast majority of gardens were located beside rivers: 65 per cent of the total area under cultivation was located in the Bitoi delta (Wagner 2002; Bein et al. 2007) and 30 per cent was located beside the Tabale River near the village. Most other gardens were located on Lababia Island (Wagner 2002). The Bitoi delta is highly valued as agricultural land for a number of reasons. Soils are regularly enriched by the siltation carried by occasional floods; fallow periods are shorter than in other gardened areas because of high fertility and the trees that grow during fallow periods are therefore smaller and easier to clear away in preparation for planting; and land is flat and therefore easier to clear, plant, weed and harvest. Close proximity to the river also provides people with ready access to drinking water during the day and access to water for bathing and washing produce, especially sweet potatoes, before beginning the long trek home at the end of the day. On some occasions, families whose gardens were located furthest upriver would transport food by outrigger canoe to the mouth of the river, thus saving themselves, and especially the women who do most of the food carrying, a great deal of hard labour. My mapping of garden locations throughout the delta also made it clear that gardens clustered in certain areas, confirming what many people had already told me—that they liked to locate their gardens close to those of relatives and friends in order to maximise socialisation time and the sharing of labour.

In June 2000, nine months after I completed my doctoral research, the Bitoi delta flooded to such an extent that virtually all gardens were destroyed. Minor flooding had occurred in April 1998, four months before my arrival, and I was aware that many households had not yet entirely recovered from their losses by the time of my arrival. When planted crops are flooded they mostly rot and die, and it is not possible to replant in that area for another four to six months. Since it also takes four to six months for staple crops like sweet potatoes to mature after they are planted, and still longer for taro, a flood can significantly reduce household food production for as long as 12 months. Although minor floods cause significant hardship, villagers have many strategies for dealing with temporary food shortages: they typically plant several gardens in different locations so that if one floods another will still be productive; they harvest more wild food crops, especially sago, to make up for shortages of sweet potato and taro; and they rely on relatives whose gardens were not damaged and to whom they provided similar help in the past. Villagers I spoke to in 2002, when I next returned to the village, told me that the flood of 2000 was beyond the experience of any living member of their community. Neither the villagers nor the scientists who worked on occasion for VDT were able to offer a definitive explanation for why flooding had been so severe in that particular year. Perhaps it was a combination of upstream logging, climate change, heavier than usual rains and more intensive farming of the delta itself that causes destabilisation of river beds due to the depletion of secondary forest along the banks. Significant minor flooding occurred twice more in the 2000s, a significant departure from the once every five to 10-year flood events that villagers reported experiencing over the previous several decades. The river thus brings significant benefits but also risks. I was not able to quantify the impact of this series of floods on gardening practices, but the anecdotal evidence I gathered suggests that the delta remained the site of intensive gardening activities throughout this decade despite these regular setbacks.

Two senior kinship groups in the village, the Amboli and Duweli, are recognised as the dominant landholding groups in the Bitoi delta (see Figure 9.2), with claims that were established well before the colonial period that began in the late 1800s. Both belong to the Gala moiety and both have distinct histories that link them to specific regions within the delta and to one of the two arms of the river. Amboli identity is closely tied to the cultivation of taro in areas near the Aleta, the northern branch of the river. Taro cultivation had virtually ceased by 1998 due to the spread of taro beetle to this region several decades previously, and discussions of taro growing at that time evoked particularly nostalgic memories of a time when taro, not sweet potato, was the dominant and most highly valued food crop. Once a highly effective beetle insecticide was made available, clan members quickly returned to taro as a preferred crop, and when I visited the village in 2013 it was abundant once again.

Yaling Michael,[7] an especially knowledgeable member of Amboli, recounted stories to me in 1999 concerning the history of movement of his ancestors from various locations near or along the Aleta River, beginning with their occupation of a village site located on a hilltop called Mundiodo, about 2 km upstream from the mouth of the river. He could trace his own ancestry back eight generations, and stated that the first occupation of Mundiodo occurred seven generations ago. The following account of this occupation comes from a story Michael copied word

---

7   All personal names appearing in this chapter are used with the knowledge and consent of the individuals involved.

for word from a book written by his father in Tok Pisin. The full story provides a partial explanation for the origin of the Balekatu, the Amboli sub-group to which he belongs:

> Long ago, during the time of our ancestors, all the Amboli lived near Aleta on top of a little mountain called Mundiodo. They built a men's house for themselves and they called this Lumi Ambo.[8] They carved all kinds of images on the house posts. After some generations our ancestors left Mt Mundiodo, beside Aleta, and they built a new men's house called Lumi Ambo on Lababia Island. Later on they moved again, building their men's house in a bay [near Aleta] called Balekatu.

The movement from Mundiodo to Lababia Island occurred as a result of a breakdown of the normally friendly relations between the Kala and a neighbouring inland group, the Biangai, who today live in several villages located near the town of Wau. After the Kala moved to Lababia Island, however, another neighbouring group, the Kaiwa, began to move into the Aleta area, giving rise to another series of violent skirmishes and an historic enmity that continues to this day. The performances of a village theatre group in 1998 emphasised the fact that, historically, men would accompany their wives in armed groups when travelling to gardens in the delta because of the ever-present possibility of Kaiwa attacks (Wagner 2002). The history of conflict between the two groups appeared to be drawing to a close by 1998, when I was first in the village, but erupted again in 2008 and dramatically shifted gardening practices for both Kamu Yali and Salus, a Kaiwa community located just north of Aleta. Many people from both communities were planting gardens in the delta in 1998, but the Kaiwa mainly planted on sites located on the northern side of Aleta or sites several kilometres inland from the mouth of the river. Once beyond 10 km from the river mouth, the delta gives way to higher ground, as one then enters the historic 'bush' territory of the Kaiwa. According to the Kala, the Kaiwa historically occupied inland forested areas, but after World War II, an Australian *kiap* (district official) moved a few Kaiwa communities down into shoreline villages in order to make it easier to reach them and provide services. According to Kamu Yali spokespeople, a Kala *tultul* (an appointed assistant to the *kiap*) gave permission to one inland Kaiwa community to move to a shoreline location just north of Aleta, and their descendants today live in the community of Salus, which

---

8   While 'clans' (*klan*) are designated in the Kala language by adding the suffix *li* to the group's core name, as with Amboli, Duweli and Aneli, men's houses are designated by placing the word *lumi* before the name, as with Lumi Ambo, Lumi Duwe or Lumi Ane.

is located slightly north of their first village site. The residents of Salus tell a somewhat different version of this story, of course, and conflicts sometimes occur. For instance, a few Kala men told me stories of times when they had cut down trees planted in their Bitoi gardens by the Kaiwa in order to prevent the latter from establishing permanent claims to those areas. They had also destroyed small dams on the Aleta that Kaiwa men had built to create fish habitat.

Between 1950, when the village of Salus came into being, and 1998, when I arrived, there had been only one instance of large-scale violence between the two communities. Village men in Kamu Yali frequently told me stories of that fight, humorously pointing out men in the village who were now quite elderly, but who had been sent to jail for one month as punishment by the Australian colonial authorities. Apparently, men from Kamu Yali and Salus had confronted one another in force on a beach just north of the mouth of the Aleta. Many individuals had been injured but no one had been killed, and few of the combatants on either side had carried lethal weapons such as bush knives, spears or guns. Several Kala men reported to me in 1997 that they had Kaiwa 'trading partners' in Salus, and there were a few other examples of peaceful ties between the two communities. There was a decided lack of intermarriage, however, and a strong sense in Kamu Yali that such ties should not occur because of the simmering land dispute. It was not entirely unexpected, then, that hostilities broke out again in 2008 following a series of disputes in the gardening areas. The intensity of what happened in 2008 was shocking and unexpected, however, to people throughout the region and to most Kamu Yali and Salus villagers themselves. According to the accounts I heard, a gang of young Salus men, one of whom carried a gun and all of whom used drugs and were involved in the drug trade, had begun to threaten and harass Kamu Yali men and women in their gardens. This behaviour was reported to the police, who tried but failed to arrest gang members. The police then suggested to several Kamu Yali men that they should take matters into their own hands and, as reported in a national newspaper (Anon. 2008a, 2008b) and confirmed for me later by villagers, they burned the entire village of Salus to the ground. Two Salus men were killed during the conflict, but, despite a prolonged police investigation, no one was ever charged for the deaths. Various attempts were made over a period of several years to deal with the issue through compensation payments, and these efforts at reconciliation eventually restored peace but have not led to a firm agreement over land rights. Salus residents have returned and rebuilt their community with Kamu Yali consent.

The Duweli are the other Gala clan with deep historic ties to the Bitoi area, but in their case the ties are to the southern branch of the river, the Daunewa (see Figure 9.4). The Duweli also cultivate taro but were more specifically known for their expertise in hunting pigs. According to the Duweli headmen whom I was able to interview as a group, pig hunting was then a collective enterprise involving the use of large nets. Today, by contrast, men hunt pigs individually with five or six dogs and a spear. Knowledge of the former Duweli pig-hunting practices is mostly lost today, but according to Duweli headmen, specialisations such as pig hunting were essential to the ways in which different Kala kin groups constructed their relationships to one another, each having special knowledge that would inform the ways in which food was procured and exchanged during the ceremonial events, such as marriages, that brought the groups together as a single community. Each group formerly maintained its own small village, centred on a men's house, which would be surrounded by the smaller houses occupied by wives and children.

Figure 9.4 Fording the Bitoi River where it divides into Aleta and Daunawe branches.

Note: The small hill in the background burned during a dry spell in 1997, the year of an El Niño episode.

Source: Photo by the author.

Accounts by Gala headmen of the origins and histories of their various clans suggest that they maintained a clear sense of alliance and solidarity amongst themselves that was less fully shared with the Aleme whose village site was located near the mouth of the Tabale River. However, although the early histories and subsistence strategies of Gala and Aleme were significantly different, those differences also appear to have facilitated well-established, long-term reciprocal relations involving the sharing and exchange of foods, resources and marriage partners. They were also firm allies against the Kaiwa.

The Alewili

The Alewili River is very different from the Tabale and Bitoi by virtue of its more rapid descent from the mountains. There is a large swamp on the northern side of the river mouth but no flood plain to support agriculture on the scale that is practised in the Bitoi delta, and far less sago than along the Tabale. Gardening is possible on the higher ground surrounding the swamp and upstream along the river for a distance of about half a kilometre, after which the steep elevation and rugged terrain make gardening here less attractive than in other locations (see Figure 9.5). This helps explain why no village site has ever been located beside the Alewili, but other factors have also influenced this choice.

Figure 9.5 Heading upstream from the mouth of the Alewili River.
Source: Photo by the author.

Gabu, one of the main leaders of the Tesiaolĩ, the clan most strongly associated with this area, explained to me in 2013 that the relationship of the Kala to this area is historically more recent than their relationship to the Bitoi and Tabale river areas. Kala people first began to frequent this area five generations ago, shortly before missionaries arrived in the area in the early 1900s. The main village site on Lababia Island had become overcrowded and families began to move to various locations near the Tabale River. Others explored areas further south and, according to Gabu, his ancestor Gõyu first began to live on Yawame Island (also known as German Island) at this time. Gõyu's father, Gosiligu, had already established a presence on the island, which is just offshore from the mouth of the Alewili, but Gõyu was the first to build a house there and plant a garden. Gõyu was very cautious in his first visits to the Alewili area on the mainland, which was previously unknown territory, both out of fear of attack by Kaiwa hunters and out of fear of the powerful spirits that resided there. He would spend short amounts of time there at first, until he felt confident that his presence was accepted, and then he began to regularly hunt and fish in the area, sleeping overnight on occasion, but still maintaining his house site on Yawame Island. This pattern continued for several years, but eventually, after missionaries arrived and built a church on Lababia Island, he converted and moved back to his previous home. His son Bolulu, however, continued to come to Yawame to garden and to the Alewili to fish and hunt. In this way, the family acquired and maintained the landownership rights they still hold over this area. Bolulu was Gabu's grandfather.

I acquired my first knowledge of this river through stories about groups of village men going there to fish in order to raise a bit of cash for the maintenance of school buildings, the church, medical aid post or for other collective village projects. Family groups would also go there to raise money for funeral expenses. In addition to its value as a fishing site, the Alewili was also one of the few places where a certain species of tree, necessary for the construction of outrigger canoes, could be found. My first visit to the Alewili occurred in 1999 during the time I was conducting a series of fishing trials with village men. Each man would take me by canoe to one or more of his favourite fishing spots where we would each fish, and I would record the size, number and species of fish caught and jot down accounts of fishing lore and sea tenure practices. Preferred fishing spots were usually located along the edges of fringing reef about half a kilometre from shore, and the Alewili was a preferred area for many. Fishing spots closer to the village tended to be less productive or more fished out, and most men did not like to go much further than Alewili on a one-day trip since it took a couple of hours of steady rowing to get there and a couple more to come back.

I was fortunate on one of my trips to be able to go with Bing Katu, then headman of the Aleme moiety. During the trip, he explained the 'rules' of sea tenure to me as they applied to fishing spots and the mouths of rivers. When he was a young man, he stated, people paid much closer attention to traditional rules, whereas now everyone in the village felt free to fish anywhere in Kamu Yali territory. He himself continued to fish only in Aleme territory, which extended from the Tabale River south to the Saia. A few fishing spots within this territory would once have been recognised as belonging to one family group, or one clan, or simply as the favourite fishing sites of one or two men—preferences that were mostly respected by other men. The Aleme headman explained to me that the leaders of specific kin groups were considered the *wasman* ('watchmen') or custodians of river mouths, and that these rights were still active despite the recent movement towards a village fishing commons as opposed to family or clan commons. He identified members of the Tesiaolî clan as custodians of the mouth of the Alewili and surrounding territory, but without mentioning their clan membership. In fact, the Tesiaolî had not even been mentioned when I first began gathering the names of village clans. Individuals who I subsequently came to understand were Tesiaolî had identified themselves to me as Mambali, a minor clan with overlapping membership, but an entirely different history. The emergence of Tesiaolî and seeming disappearance of Mambali during the 1999–2013 period appears almost entirely attributable to villagers' responses to mining exploration.

The Alewili today is still regarded as the home of dangerous and powerful spirits, and villagers do not come here unless they are Tesiaolî or are accompanied by someone who is. My first experience of the area, in 1999, provided me with a striking confirmation of the special relationship between the Alewili and the Tesiaolî. Towards the end of my doctoral research year, Yaling Michael invited me to go there on an overnight 'picnic', something I rarely experienced in the village, and I happily agreed. Since Michael has no clan rights in the area he invited Yaga, a Tesiaolî member, to come with us. Michael maintained a close relationship with Yaga by virtue of the fact that they were married to sisters and often worked together in their gardens. This first trip was especially memorable because of heavy rains that fell in the night and threatened to wash away our small campsite, where I had pitched my tent and where the others had constructed small, temporary sleeping huts for themselves. In the morning, I was told that Yaga had arisen in the night and, noticing that the river was beginning to overflow its banks, had quickly cut down several nearby saplings and created a row of stakes between our camp and the rising water. He then performed a ritual in which he asked the water to come no further than the line of stakes. When I was told this story in the morning, everyone pointed out to me, with a good deal of satisfaction but also amusement, that the river had risen exactly to the line of stakes and then retreated, as demonstrated by the wetness of the sand on the river side of the stakes.

This was one of the very few occasions during my year-long stay in the village in 1998–99 when villagers openly acknowledged or demonstrated their continuing use of Indigenous, non-Christian, ritual knowledge. My first such experience had occurred when I went on an overnight trip into the mountains with three hunters. These hunters were all well-known to me, and all had described hunting practices to me before, but once in the bush, sitting at night around a fire, they began to describe some of the ways in which their hunting activities were guided by the kinds of dreams and magical knowledge so often described in the PNG literature on this topic. Previous attempts to question villagers directly on the use of non-Christian ritual, in gardens or when fishing or hunting, had always met with denials that any such practices existed, though, on some occasions, stories were told of such practices and regret was expressed at the loss of knowledge that had occurred since adopting Christian practices. The ritual line of stakes in the ground at Alewili thus reinforced my sense that while Christianity was clearly the dominant religion in the village, Indigenous practices continued to hold sway in the bush. Villagers have abandoned and forgotten many—perhaps most—Indigenous practices, but under certain circumstances they are remembered and revived under the guidance of those who did not forget.

Mineral analyses conducted by the Katana mining company in 2012 confirmed that the area surrounding the mouth of the Alewili River contains a rich chromite deposit. The company also confirmed the findings of previous exploration companies: that more extensive but less rich deposits were present at the mouth of the Saia River. As villagers began to mobilise in response to the Katana initiative, the underlying logic of their land tenure system was revealed with singular clarity. Those who hold the strongest landownership claims to Alewili are the descendants of Gõyu. At the time when he moved south to Yawame Island and to the mouth of the Alewili River, Gõyu belonged to the Balekatu, a Gala clan closely related to the Amboli, but his claim to the Alewili does not extend to the Balekatu clan as a whole; rather, it remains with him and his own immediate descendants. His personal claim to Alewili is based on the fact that he was the first to fish, hunt and garden in the area—activities that all required him to cultivate a personal relationship with the spirits of the place and gain their acceptance. This appears to be the most fundamental of all the underlying principles that inform landownership practices at Kamu Yali. As a primary principle, it generates almost endless opportunities for the collapse and re-configuration of landowning groups in response to the changing values of particular resources. Now that the Alewili has acquired new value as a chromite mining site, all descendants of Gõyu, whether traced through male or female lines or both, have reason and opportunity to declare their primary affiliation as Tesiaolĩ. The size and status of Tesiaolĩ has thus expanded at the expense of other kin groups, most notably Mambali.

The Saia

According to Thomas Yawing, a leading member of the Ane clan, the main landowning group in the Saia area, an ancestor by the name of Kapalia moved south from Kamu Yali in the late 1800s, around the same time as Gõyu. He went first to Yawame Island, then to Lasanga Island near present-day Kui, as a result of an epidemic that struck Kamu Yali in the late 1800s. Like the Alewili, the Saia area at that time had a reputation as a place of dangerous spirits, but also as a place of abundant resources. The most dangerous spirits are said to inhabit a swampy area adjacent to the river itself. There are areas suitable for gardening but they are not extensive. Wild game, fish, turtles and wild sago are abundant, and also a particular species of pandanus tree valued for its fruit. The clan name, Ane, is, in fact, derived from the name of this particular tree.

According to Thomas and Hamm Giamsa, another member of Aneli, Kapalia and his sons gradually built a relationship with the spirit beings who occupied the region. Thomas and Hamm also related a story about how Kobõmbia, one of Kapalia's sons, led a group of Kala men in battle against a Kaiwa group that came to challenge their claim to this area. Their victory in battle thus consolidated their claim. Thomas and Hamm emphasised the reputation of Aneli at that time as powerful sorcerers who

243

augmented their power by eating their fallen victims. They also stated that a collection of material items dating from around this time, such as knives, shell jewellery and other decorative materials, were collected and hidden in a secret location in the forest to preserve a record of their historic occupation of the area. This site remained intact until the close of World War II, when allied US and Australian forces used aerial bombardment when retaking this area from the occupying Japanese. No one has since been able to locate the site.

When I was first in Kamu Yali in 1998, I was told that a local court ruling in the 1970s had declared the Saia River to be the boundary between Kamu Yali and Buso, and that the entire area on the Kamu Yali side fell under the authority of the Aleme moiety. The court case arose in response to chromium exploration activities in the area by Conzinc Riotinto of Australia, the first company to conduct such studies. But Thomas and Hamm did not even mention this court case when they recounted Aneli history for me in 2013, and they did not indicate that the Aleme were senior holders of rights to the Saia area. This did not entirely surprise me, since it was consistent with other changes in the ways villagers were describing kin structures in 2013 by comparison to 1998. The local land court had apparently ruled that the Saia River was the boundary between the two villages on the basis of an argument put forward by the headmen of the Aleme moiety at Kamu Yali. The headmen argued that the rights of the Aneli, since they are an Aleme clan, are subject to the overarching authority of the moiety, or 'major clan', and that the Saia therefore falls under Aleme control. The land court magistrate who ruled on the case accepted this argument, but because Aneli spokespeople from Buso disputed the Aleme claim, he felt compelled to establish village boundaries as well. Unfortunately, he did so in a way that entirely confounds the historic relationship of the Aneli in Buso to the Aneli in Kamu Yali. The Aneli have always been the largest clan at Buso, though other clans such as the Duweli and Amboli are also present. But, unlike the situation at Kamu Yali, where everyone belongs to one of two moieties, either Gala or Aleme, everyone at Buso belongs to a single group known as Bwaboli, a name that derives from Bwabo, the place on Lasanga Island where Kapalia and his family lived before establishing themselves at Saia and founding the community of Buso. The Aneli at Buso therefore do not consider themselves subject to the higher authority of the Aleme moiety at Kamu Yali.

This is only one of several instances in which the prominence of the Aleme moiety has diminished over the past decade in respect to land rights, but rather than overwhelm readers with more details about the fluid nature of Kala kinship structures, I wish to return to the question of memory and forgetting in respect to rivers and the passage of time. The individual who first told me about the 1970s court case also showed me a two-page fragment of a typed copy of a court decision that he had obtained from his father, one of the Aleme leaders at the time, long ago deceased. When I subsequently tried to locate a copy of the full, written court decision from the provincial Department of Lands, I was told, after repeated visits and an exhaustive search, that they had no such record in their possession. His short fragment, yellowed with time and edged with mildew, about to pass out of collective memory altogether, thus provides an unexpected metaphor for the mutability of human actions as they erode with the passage of time and the capacity of each generation to select and discard the fragments that hold most meaning in relation to contemporary events.

As the court case disappears from memory, other bodies of knowledge about the Saia are recovered and reassembled, vital once again despite their suppression by missionaries over 100 years ago. Spokespeople and knowledge keepers for both the Aneli and Tesiaolī clans now find themselves evaluating the value of that knowledge by comparison to the economic value of chromite mining. Those I have spoken to express deeply conflicted emotions when asked whether they will agree to allow mining at the mouths of their rivers. Thomas Yawing, for instance, states that he is opposed to mining because of his worries about environmental damage, and because mining activities will obliterate the relationships he has nurtured with the ancestor spirits of the Saia area during his lifetime. But, according to Thomas, the general consensus among Aneli members may be in favour of mining. Hamm, on the other hand, after seeing half a dozen mining exploration companies come and go since the late 1960s, does not believe that actual mining will ever occur, and therefore sees no reason to take a side one way or the other.

Conclusion

Rivers shape and transform the land over which they flow and the seas into which they empty; they sustain an endless number of unique riparian habitats and the lives of the many species that thrive in their immediate vicinity. Human beings are one such species, and coastal people, no less than inland people, are almost always found near rivers, creeks or other easily accessible sources of fresh water. The human need for fresh water continues to dominate and predict settlement patterns everywhere in the world and, despite our hydro-engineering prowess, desert cities like Las Vegas and Phoenix in the United States, cities that long ago outgrew their local water supplies, are still the exception rather than the rule. We can, usually at great cost to ourselves as well as other species, change the path of flowing water, capture it in concrete channels, put it to work as an agent of industrial development—but these actions emphasise rather than diminish the power of water to shape social lives.

The Kala people cannot unambiguously be categorized as a 'river people', but they can be described as a 'river mouth people', and this is true not only of Kamu Yali but of all six Kala villages, all located on the beaches of the Solomon Sea near the mouths of rivers. Their strong dependence on rivers, despite their close proximity to the sea, may be an outcome of the lower productivity of their marine environment by comparison to the Marovo Lagoon described by Hviding (this volume). But as both Mondragón (this volume) and Hviding emphasise, river mouths are where fresh water and salt water meet, where land and sea intermingle, and as a consequence are unusually rich and diverse environments, equally potent in social, ecological and cosmological terms.

The social history of Kamu Yali is, in large measure, the outcome of its relationships with the four distinct rivers described in this chapter. Villagers' relations with non-Kala neighbours are structured around their willingness to go to war, if necessary, to defend their river adaptations. Intra-village relations are structured around the distinct practices and identities of the kin groups associated with each river, distinctions that arose historically by virtue of the fact that one river environment was more suitable for growing taro, another for harvesting sago, another for fishing, another for gathering the fruit of the pandanus palm. I have also demonstrated in this chapter how a new kin group, such as the Tesiaolĩ, can emerge from the actions of an individual or family group who, through a combination of need, opportunity, courage and desire, come to know a river for the first time or come to know it in a new way.

Like the Sepik River described by Silverman (this volume), Kamu Yali rivers have so far resisted the development aspirations of their human population. The Bitoi delta has not become a site for commercial agriculture largely because of the high cost of transporting produce by boat to regional or more distant markets. The Tabale River's swamps supply villagers with a ready supply of sago and are home to a growing population of crocodiles, but crocodile tours, though attempted, have never featured prominently in the modest ecotourism industry Kamu Yali has been able to sustain over the past decade and a half. Mining activities at the mouths of the Alewili and Saia rivers, however, now present villagers with a significant development opportunity. Chromite exploration at the mouths of these rivers has tantalised villagers since the 1960s with the possibility of royalty income and wage labour, but no mining has yet taken place. Unlike previous exploration companies, however, Katana has formally applied to the PNG Government for a mining licence, and in 2013 they conducted the impact assessment studies necessary to obtain such a licence. Today the Tesiaolĩ, Aneli and their near relatives are remembering and reconstructing largely forgotten bodies of knowledge concerning the special relationships of their ancestors Gõyu and Kapulia to the Alewili and Saia; they must do this in order to confirm their claims as landowners. At the same time, however, they are being asked to forget, to throwaway the practices through which their ancestors constructed their relationships with these rivers, to dismiss them as of less value than the financial rewards of mining and the dream of a 'modern' life.

In their own way, the rivers are also engaged in a complex process of memory and forgetting. The chromite deposits present at the mouth of the Alewili and Saia are alluvial deposits; their concentration at these sites is largely the outcome of the actions of the rivers over thousands of years. Chromium extraction, if it occurs, will entirely rearrange these formations, degrading both downstream riparian zones and the adjacent marine environments. But the rivers will continue to carry water nonetheless from Batatalã to the sea, will continue to arrange and rearrange the sediments at their mouths, forgetting each old arrangement as a new one emerges. Without losing sight of the mnemonic significance of rivers, as emphasised by Harrison (2004), I would like to conclude by also noting their mimetic significance. As Krause (2010), Edgeworth (2011, 2014), Féaux de la Croix (2011) and Strang (2014) propose, and as noted in the introduction to this volume, the flow of water provides a useful metaphor for both the passage of time and the fluidity of cultural forms as they respond over time to social and ecological actions and reactions. Harrison (2004: 141) evokes the mimetic as well as mnemonic qualities of rivers when he describes the broad, marshy plains of the meandering Sepik River as a 'forgetful landscape'. From this perspective, culture itself is very like a river, and cultural analysis gains traction through careful observation of river ecology and a willingness to treat rivers, as well as humans, as the subject of ethnographic enquiry.

Acknowledgments

Funding for the fieldwork that informs this paper was provided by the Social Sciences and Humanities Research Council of Canada, the International Development Research Centre of Canada, Okanagan University College, the University of British Columbia Okanagan, the University of British Columbia Hampton Foundation, the Firebird Foundation for Anthropological Research and the US National Science Foundation. I would like to thank all those who participated in the conference sessions that led to this publication and the Association for Social Anthropology in Oceania for sponsoring the sessions. I would also like to acknowledge the insightful and constructive comments of two anonymous reviewers. Finally, I would also like to express my deep gratitude to the many Kamu Yali residents who have shared their time, insights and hospitality with me over a period of two decades now, especially those whose knowledge and experience most directly informed this paper.

References

Anon., 2008a. 'Investigators Dig-Up Evidence.' *Post-Courier*, 18 January.

——, 2008b. 'Police Role in Fatal PNG Raids Denied.' *Post-Courier*, 21 January.

Barad, K., 2003. 'Posthumanist Performativity: Toward an Understanding of How Matter Comes to Matter.' *Signs: Journal of Women in Culture and Society* 28: 801–831. doi.org/10.1086/345321

Bein, F.L. (ed.), 1998. 'Kamiali Wildlife Management Area Bio-Diversity Inventory: A Report Prepared for Village Development Trust.' Lae: PNG University of Technology, Environmental Research and Management Centre.

Bein, F.L., J. Wagner and J. Wilson. 2007. 'Food Garden Carrying Capacity of the Bitoi River Delta for the Kamiali Wildlife Management Area.' *Focus on Geography* 50(2): 28–33. doi.org/10.1111/j.1949-8535.2007.tb00194.x

Bennett, J., 2010. *Vibrant Matter: A Political Ecology of Things*. Durham (NC): Duke University Press.

Burton, J., 1996. 'Aspects of Biangi Society: The Solorik System.' Canberra: Pacific Social Mapping (Hidden Valley Project Working Paper 3).

Descola, P., 2013. *Beyond Nature and Culture* (transl. J. Lloyd). Chicago: University of Chicago Press.

Edgeworth, M., 2011. *Fluid Pasts: Archaeology of Flow*. London: Bloomsbury.

——, 2014. 'On the Agency of Rivers.' *Archaeological Dialogues* 21: 157–159. doi.org/10.1017/S1380203814000166

Féaux de la Croix, J., 2011. 'Moving Metaphors We Live By: Water and Flow in the Social Sciences and around Hydroelectric Dams in Kyrgyzstan.' *Central Asian Survey* 30: 487–502. doi.org/10.1080/02634937.2011.614097

Harrison, S., 2004. 'Forgetful and Memorious Landscapes.' *Social Anthropology* 12: 135–151. doi.org/10.1017/S0964028204000436

Hodder, I., 2012. *Entangled: An Archaeology of the Relationships between Humans and Things*. New York: Wiley. doi.org/10.1002/9781118241912

Ingold, T., 2000. *The Perception of the Environment: Essays on Livelihood, Dwelling and Skill*. London: Routledge. doi.org/10.4324/9780203466025

Ingram, M., 2010. 'Fermentation, Rot, and Other Human-Microbial Performances.' In M. Goldman, P. Nadasdy and M. Turner (eds), *Knowing Nature: Conversations at the Intersection of Political Ecology and Science Studies*. Chicago: University of Chicago Press.

Krause, F., 2010. Thinking Like a River: An Anthropology of Water and its Uses along the Kemi River, Northern Finland. Aberdeen: University of Aberdeen (PhD thesis).

Latour, B., 2005. *Reassembling the Social: An Introduction to Actor-Network Theory*. Oxford: Oxford University Press.

Malafouris, L. and C. Renfrew, 2013. *How Things Shape the Mind*. Cambridge: MIT Press.

McKellin, W.H., 1991. 'Hegemony and the Language of Change: The Pidginization of Land Tenure among the Managalase of Papua New Guinea.' *Ethnology* 30: 313–324. doi.org/10.2307/3773687

Morton, T., 2013. *Hyperobjects: Philosophy and Ecology after the End of the World*. Minneapolis: University of Minnesota Press.

Schreyer, C. and J. Wagner, 2013. 'Kala Biŋatuwã: A Community-Driven Alphabet for the Kala Language.' In N. Ostler and M. Norris (eds), *Endangered Languages Beyond Boundaries: Community Connections, Collaborative Approaches, and Cross-Disciplinary Research*. Ottawa: University of Carleton Press.

Strang, V., 2014. 'Fluid Consistencies: Material Relationality in Human Engagements.' *Archaeological Dialogues* 21: 133–150. doi.org/10.1017/S1380203814000130

Viveiros de Castro, E., 1998. 'Cosmological Deixis and Amerindian Perspectivism'. *Journal of the Royal Anthropological Institute* 4(3): 469–488.

Wagner, J., 1999. '"Blue Mountains Constantly Walking": The Resignification of Nature and the Reconfiguration of the Commons in Rural Papua New Guinea.' Canberra: The Australian National University, Resource Management in Asia-Pacific Program (Working Paper 24).

——, 2002. Commons in Transition: An Analysis of Social and Ecological Change in a Coastal Rainforest Environment in Rural Papua New Guinea. Montreal: McGill University (PhD thesis).

——, 2005. 'The Politics of Accountability: An Institutional Analysis of the Conservation Movement in Papua New Guinea.' In P.J. Stewart and A. Strathern (eds), *Anthropology and Consultancy: Issues and Debates*. New York: Berghahn Books.

——, 2007. 'Conservation as Development in Papua New Guinea: The View from Blue Mountain.' *Human Organization* 66: 28–37. doi.org/10.17730/humo.66.1.q21q23v06t374204